高等院校计算机应用系列教材

计算机网络实验实践教程

王鸿运　韩　芳　郑晶晶　主　编

清华大学出版社
北京

内 容 简 介

本书旨在帮助读者在学习计算机网络基础知识之后，进行网络设备操作、配置、设计和综合应用的上机实战训练。内容涵盖了常用交换路由硬件设备知识及交换机与路由器的详细配置。每个实验均给出实验目的、实验要求、实验拓扑、实验设备、实验步骤等，并且安排了一定数量的上机实战操作，能够让读者通过实验操作和实战练习来巩固相关的知识并掌握技能。

本书内容翔实，具有很强的实用性，着重动手能力的培养。书中每一个实训都经过精心挑选，并在操作步骤上给予了详细说明。

本书可作为本专科院校计算机类专业、电子信息类专业、通信类专业、电子商务类专业及其他专业的计算机网络课程配套实验实训教材，也可作为大中专院校、成人高校、计算机培训机构的实验实训指导用书。

本书配套的电子课件、教学大纲和习题答案可以到 http://www.tupwk.com.cn/downpage 网站下载，也可以通过扫描前言中的二维码下载，读者扫描前言中的教学视频二维码可以观看学习视频。

本书封面贴有清华大学出版社防伪标签，无标签者不得销售。
版权所有，侵权必究。举报：010-62782989，beiqinquan@tup.tsinghua.edu.cn

图书在版编目(CIP)数据

计算机网络实验实践教程 / 王鸿运，韩芳，郑晶晶
主编. -- 北京：清华大学出版社, 2025. 1. -- (高等
院校计算机应用系列教材). -- ISBN 978-7-302-67736-9
Ⅰ. TP393-33
中国国家版本馆 CIP 数据核字第 2024DD7540 号

责任编辑：胡辰浩
封面设计：高娟妮
版式设计：妙思品位
责任校对：孔祥亮
责任印制：刘海龙

出版发行：清华大学出版社
    网　　址：https://www.tup.com.cn，https://www.wqxuetang.com
    地　　址：北京清华大学学研大厦 A 座　　邮　　编：100084
    社 总 机：010-83470000　　　　　　　　邮　　购：010-62786544
    投稿与读者服务：010-62776969，c-service@tup.tsinghua.edu.cn
    质 量 反 馈：010-62772015，zhiliang@tup.tsinghua.edu.cn
印 装 者：三河市天利华印刷装订有限公司
经　　销：全国新华书店
开　　本：185mm×260mm　　　　印　　张：18.5　　　　字　　数：439 千字
版　　次：2025 年 1 月第 1 版　　　　　　　　　　　　印　　次：2025 年 1 月第 1 次印刷
定　　价：79.00 元

产品编号：109362-01

前　言

在当今信息爆炸的时代，互联网已成为全球一体化的重要纽带，计算机网络技术深入渗透到人类生活的每一个角落，被誉为"近代最深刻的技术革命"。计算机网络不仅推动了社会信息化的进程，也极大地促进了经济的发展。"网络时代"和"网络经济"等词汇生动地描述了计算机网络在现代社会中的重要性。社会的信息化、数据的分布式处理和计算机资源的共享等需求，持续推动着计算机网络技术的快速发展。

计算机网络是计算机技术与通信技术密切结合的学科，也是计算机应用中一个空前活跃的领域。本课程不仅是计算机科学与技术专业的主干课程，也是电子与通信专业学生以及广大从事计算机应用和信息管理的科技人员都必须学习的课程。同时，我国的信息化建设也需要大量掌握计算机网络基础知识和应用技术的专业人才。本课程不仅是一门理论性很强的课程，同时也是一门实践性很强的课程。因此，为了配合计算机网络的实验教学，我们以《计算机网络技术》为背景，结合实验室的设备和仪器编写了这本实验指导书。以帮助学生通过严格的实践训练真正掌握和深入理解计算机网络的基本理论，协议和算法。

本书是面向理工科的本专科学生的计算机网络实验指导教材。结合理论数学精心设计了交换机实验17个、路由器实验15个、实训项目11个和网络综合实践6个，有利于学生在掌握一定理论和实践之后进行设计和创新。

本书在实验内容组织上具有较强的系统性和可操作性，所要求的实验环境相对简单，所有实验内容均在实验室完成。学生通过完成设计的实验内容能够深入掌握和理解计算机网络的工作原理和工作过程，增强处理实际问题的能力。

为了优化实验教学，建议学生每两人组成一个小组，每组配备两台计算机，其中一台作为服务器，另一台作为客户机。网络配置实验可以根据实际需求进行设置，最终实现访问功能。实验所需的工具、软件和材料基本为每组分配一套，而有些工具，如网线测试仪等，可以由各组共同使用。

本书可作为本专科院校计算机类专业、电子信息类专业、通信类专业、电子商务类专业及其他专业的计算机网络课程配套实验实训教材，也可作为大中专院校、成人高校、计算机培训机构的实验实训指导用书。

本书由黄河科技学院教师王鸿运、韩芳、郑晶晶主编，期间得到民办教育品牌专业建设-计算机科学与技术(项目编号：ZLG201903)、郑州地方高校急(特)需专业-数据科学与大数据技术(项目编号：ZZLG202301)和河南省研究生教育改革与质量提升工程项目(项目号：YJS2023JD67)资助。由于编者水平有限，书中难免有不足之处，欢迎广大读者批评指正。在编写本书的过程中参考了相关文献，在此向这些文献的作者深表感谢。我们的电话是010-62796045，邮箱是 992116@qq.com。

  本书配套的电子课件、教学大纲和习题答案可以到 http://www.tupwk.com.cn/downpage 网站下载。也可以扫描下方左侧的二维码获取。扫码下方右侧的二维码可以直接观看教学视频。

配套资源　　　　　　　　　　　扫一扫

扫描下载　　　　　　　　　　　看视频

编　者

2024 年 8 月

目 录

第1章 交换机实验 ··········· 1

实验一 交换机带外管理 ············ 1
 一、实验目的 ···················· 1
 二、相关知识 ···················· 1
 三、实验设备 ···················· 1
 四、实验拓扑 ···················· 2
 五、实验要求 ···················· 2
 六、实验步骤 ···················· 2
 七、课后练习 ···················· 5

实验二 交换机的配置模式 ········· 6
 一、实验目的 ···················· 6
 二、相关知识 ···················· 6
 三、实验设备 ···················· 8
 四、实验拓扑 ···················· 8
 五、实验要求 ···················· 8
 六、实验步骤 ···················· 8
 七、注意事项和排错 ········· 12
 八、思考题 ···················· 12
 九、课后练习 ·················· 12

实验三 交换机CLI界面调试技巧 ··· 13
 一、实验目的 ···················· 13
 二、应用环境 ···················· 13
 三、实验设备 ···················· 13
 四、实验拓扑 ···················· 13
 五、实验要求 ···················· 13
 六、实验步骤 ···················· 14
 七、课后练习 ···················· 16

实验四 交换机恢复出厂设置及其
    基本配置 ················ 16
 一、实验目的 ···················· 16

 二、相关知识 ···················· 16
 三、实验设备 ···················· 16
 四、实验拓扑 ···················· 17
 五、实验要求 ···················· 17
 六、实验步骤 ···················· 17
 七、注意事项和排错 ········· 20
 八、配置序列 ···················· 20
 九、思考题 ···················· 20
 十、课后练习 ·················· 20

实验五 使用Telnet方式管理交换机 ··· 21
 一、实验目的 ···················· 21
 二、相关知识 ···················· 21
 三、实验设备 ···················· 21
 四、实验拓扑 ···················· 21
 五、实验要求 ···················· 22
 六、实验步骤 ···················· 22
 七、注意事项和排错 ········· 25
 八、配置序列 ···················· 26
 九、思考题 ···················· 26
 十、课后练习 ·················· 26

实验六 使用Web方式管理交换机 ··· 26
 一、实验目的 ···················· 26
 二、相关知识 ···················· 27
 三、实验设备 ···················· 27
 四、实验拓扑 ···················· 27
 五、实验要求 ···················· 27
 六、实验步骤 ···················· 28
 七、注意事项和排错 ········· 29
 八、配置序列 ···················· 30
 九、思考题 ···················· 30
 十、课后练习 ·················· 30

实训项目一　交换机配置模式及管理…… 30
　　一、实训目的…………………… 30
　　二、实训设备…………………… 31
　　三、实训拓扑…………………… 31
　　四、实训内容…………………… 31
　　五、实训步骤…………………… 31

实验七　交换机 VLAN 划分实验…… 36
　　一、实验目的…………………… 36
　　二、相关知识…………………… 37
　　三、实验设备…………………… 37
　　四、实验拓扑…………………… 37
　　五、实验要求…………………… 38
　　六、实验步骤…………………… 38
　　七、注意事项和排错…………… 42
　　八、配置序列…………………… 42
　　九、思考题……………………… 44
　　十、课后练习…………………… 44

实验八　跨交换机相同 VLAN 间通信…… 44
　　一、实验目的…………………… 44
　　二、相关知识…………………… 44
　　三、实验设备…………………… 45
　　四、实验拓扑…………………… 45
　　五、实验要求…………………… 45
　　六、实验步骤…………………… 46
　　七、注意事项和排错…………… 52
　　八、配置序列…………………… 52
　　九、思考题……………………… 55
　　十、课后练习…………………… 55

实训项目二　组建简单交换式网络…… 55
　　一、实训目的…………………… 55
　　二、实训设备…………………… 56
　　三、实训拓扑…………………… 56
　　四、实训内容…………………… 56
　　五、实训步骤…………………… 57

实验九　交换机 MAC 与 IP 的绑定…… 59
　　一、实验目的…………………… 59
　　二、相关知识…………………… 60
　　三、实验设备…………………… 60
　　四、实验拓扑…………………… 60

　　五、实验要求…………………… 60
　　六、实验步骤…………………… 61
　　七、注意事项和排错…………… 62
　　八、课后练习…………………… 62

实验十　生成树实验…………………… 63
　　一、实验目的…………………… 63
　　二、相关知识…………………… 63
　　三、实验设备…………………… 63
　　四、实验拓扑…………………… 64
　　五、实验要求…………………… 64
　　六、实验步骤…………………… 64
　　七、注意事项和排错…………… 67
　　八、思考题……………………… 67
　　九、课后练习…………………… 67

实验十一　交换机链路聚合…………… 68
　　一、实验目的…………………… 68
　　二、相关知识…………………… 68
　　三、实验设备…………………… 68
　　四、实验拓扑…………………… 69
　　五、实验要求…………………… 69
　　六、实验步骤…………………… 69
　　七、注意事项和排错…………… 81
　　八、课后练习…………………… 81

实训项目三　基于 VLAN 的交换机
　　　　　　链路聚合……………… 82
　　一、实训目的…………………… 82
　　二、实训设备…………………… 82
　　三、实训拓扑…………………… 82
　　四、实训要求…………………… 82
　　五、实训步骤…………………… 83

实验十二　认识三层交换机…………… 83
　　一、实验目的…………………… 83
　　二、相关知识…………………… 84
　　三、实验设备…………………… 84
　　四、实验拓扑…………………… 84
　　五、实验要求…………………… 84
　　六、实验步骤…………………… 85
　　七、注意事项…………………… 88
　　八、课后练习…………………… 88

实验十三　多层交换机 VLAN 的划分
　　　　　和 VLAN 间路由 ·········· 88
　　一、实验目的 ·························· 88
　　二、相关知识 ·························· 88
　　三、实验设备 ·························· 89
　　四、实验拓扑 ·························· 89
　　五、实验要求 ·························· 89
　　六、实验步骤 ·························· 90
　　七、注意事项和排错 ·············· 97
　　八、配置序列 ·························· 97
　　九、思考题 ···························· 99
　　十、课后练习 ························ 100

实验十四　多层交换机实现二层交换机
　　　　　VLAN 间路由 ············ 100
　　一、实验目的 ························ 100
　　二、相关知识 ························ 100
　　三、实验设备 ························ 100
　　四、实验拓扑 ························ 101
　　五、实验要求 ························ 101
　　六、实验步骤 ························ 102
　　七、注意事项和排错 ············ 109
　　八、思考题 ·························· 109

实训项目四　三层交换机实现二层交换机
　　　　　　不同 VLAN 间通信 ···· 109
　　一、实训目的 ························ 109
　　二、实训设备 ························ 109
　　三、实训拓扑 ························ 110
　　四、实训要求 ························ 110
　　五、实训步骤 ························ 111

实验十五　多层交换机静态路由实验 ···· 112
　　一、实验目的 ························ 112
　　二、相关知识 ························ 112
　　三、实验设备 ························ 113
　　四、实验拓扑 ························ 113
　　五、实验要求 ························ 113
　　六、实验步骤 ························ 114
　　七、注意事项和排错 ············ 128
　　八、思考题 ·························· 128

实训项目五　多层交换机静态路由
　　　　　　配置 ························· 129
　　一、实训目的 ························ 129
　　二、实训设备 ························ 129
　　三、实训拓扑 ························ 129
　　四、实训要求 ························ 129
　　五、实训步骤 ························ 129

实验十六　三层交换机 RIP 动态路由 ···· 130
　　一、实验目的 ························ 130
　　二、相关知识 ························ 130
　　三、实验设备 ························ 131
　　四、实验拓扑 ························ 131
　　五、实验要求 ························ 131
　　六、实验步骤 ························ 132
　　七、注意事项和排错 ············ 147
　　八、思考题 ·························· 147

实训项目六　多层交换机动态 RIP 路由
　　　　　　配置 ························· 147
　　一、实训目的 ························ 147
　　二、实训设备 ························ 147
　　三、实训拓扑 ························ 148
　　四、实训要求 ························ 148
　　五、实训步骤 ························ 148

实验十七　三层交换机 OSPF 动态
　　　　　路由 ···························· 150
　　一、实验目的 ························ 150
　　二、相关知识 ························ 150
　　三、实验设备 ························ 151
　　四、实验拓扑 ························ 151
　　五、实验要求 ························ 151
　　六、实验步骤 ························ 152
　　七、注意事项和排错 ············ 165
　　八、思考题 ·························· 165

实训项目七　多层交换机之间的动态
　　　　　　OSPF 配置 ················ 166
　　一、实训目的 ························ 166
　　二、实训设备 ························ 166
　　三、实训拓扑 ························ 167

四、实训要求……………………167
　　五、实训步骤……………………167

第2章　路由器实验……………169

实验一　路由器接口简介…………169
　　一、实验目的……………………169
　　二、相关知识……………………169
　　三、实验设备……………………169
　　四、实验拓扑……………………169
　　五、实验要求……………………170
　　六、实验步骤……………………170
　　七、注意事项和排错……………170

实验二　路由器的基本管理方法……170
　　一、实验目的……………………170
　　二、相关知识……………………171
　　三、实验设备……………………171
　　四、实验拓扑……………………171
　　五、实验要求……………………171
　　六、实验步骤……………………171
　　七、注意事项和排错……………177
　　八、思考题………………………177
　　九、课后练习……………………177

实验三　路由器的基本配置…………177
　　一、实验目的……………………177
　　二、相关知识……………………177
　　三、实验设备……………………178
　　四、实验拓扑……………………178
　　五、实验要求……………………178
　　六、实验步骤……………………178
　　七、注意事项和排错……………185
　　八、配置序列……………………185
　　九、思考题………………………185
　　十、课后练习……………………185

实验四　路由器的文件维护…………185
　　一、实验目的……………………185
　　二、相关知识……………………185
　　三、实验设备……………………186
　　四、实验拓扑……………………186
　　五、实验要求……………………186
　　六、实验步骤……………………186
　　七、注意事项和排错……………189
　　八、配置序列……………………189
　　九、思考题………………………189
　　十、课后练习……………………189

实验五　单臂路由实验………………189
　　一、实验目的……………………189
　　二、相关知识……………………189
　　三、实验设备……………………190
　　四、实验拓扑……………………190
　　五、实验要求……………………190
　　六、实验步骤……………………191
　　七、课后练习……………………193

实验六　路由器静态路由配置………193
　　一、实验目的……………………193
　　二、相关知识……………………193
　　三、实验设备……………………194
　　四、实验拓扑……………………194
　　五、实验要求……………………194
　　六、实验步骤……………………194
　　七、注意事项和排错……………197
　　八、配置序列……………………197
　　九、思考题………………………198

实训项目八　多路由器间静态路由配置……………198
　　一、实训目的……………………198
　　二、实训设备……………………198
　　三、实训拓扑……………………198
　　四、实训要求……………………199

实验七　路由器 RIP-1 配置…………200
　　一、实验目的……………………200
　　二、相关知识……………………200
　　三、实验设备……………………200
　　四、实验拓扑……………………200
　　五、实验要求……………………200
　　六、实验步骤……………………201
　　七、注意事项和排错……………205
　　八、配置序列……………………205
　　九、思考题………………………206

实训项目九　多路由器间的 RIP-1 路由
　　配置 ················ 207
　　一、实训目的 ············ 207
　　二、实训设备 ············ 207
　　三、实训拓扑 ············ 207
　　四、实训要求 ············ 207
实验八　路由器 RIP-2 配置 ······ 208
　　一、实验目的 ············ 208
　　二、相关知识 ············ 208
　　三、实验设备 ············ 209
　　四、实验拓扑 ············ 209
　　五、实验要求 ············ 209
　　六、实验步骤 ············ 209
　　七、注意事项和排错 ······ 213
　　八、配置序列 ············ 213
　　九、思考题 ·············· 214
实训项目十　多路由器间的 RIP-2 路由
　　配置 ················ 214
　　一、实训目的 ············ 214
　　二、实训设备 ············ 214
　　三、实训拓扑 ············ 214
　　四、实训要求 ············ 215
实验九　静态路由和直连路由引入
　　配置 ················ 216
　　一、实验目的 ············ 216
　　二、相关知识 ············ 216
　　三、实验设备 ············ 216
　　四、实验拓扑 ············ 216
　　五、实验要求 ············ 216
　　六、实验步骤 ············ 217
　　七、注意事项和排错 ······ 219
　　八、配置序列 ············ 219
　　九、思考题 ·············· 220
　　十、课后练习 ············ 220
实验十　单区域 OSPF 基本配置 ······ 221
　　一、实验目的 ············ 221
　　二、相关知识 ············ 221
　　三、实验设备 ············ 221

　　四、实验拓扑 ············ 221
　　五、实验要求 ············ 221
　　六、实验步骤 ············ 221
　　七、注意事项和排错 ······ 225
　　八、配置序列 ············ 226
　　九、思考题 ·············· 227
　　十、课后练习 ············ 227
实验十一　RIP-2 邻居认证配置 ······ 227
　　一、实验目的 ············ 227
　　二、相关知识 ············ 228
　　三、实验设备 ············ 228
　　四、实验拓扑 ············ 228
　　五、实验要求 ············ 228
　　六、实验步骤 ············ 228
　　七、注意事项和排错 ······ 231
　　八、配置序列 ············ 231
　　九、思考题 ·············· 232
实验十二　多区域 OSPF 配置 ······ 232
　　一、实验目的 ············ 232
　　二、相关知识 ············ 233
　　三、实验设备 ············ 233
　　四、实验拓扑 ············ 233
　　五、实验要求 ············ 233
　　六、实验步骤 ············ 234
　　七、注意事项和排错 ······ 235
　　八、配置序列 ············ 235
　　九、思考题 ·············· 236
实训项目十一　多路由器间的 OSPF
　　路由配置 ············ 236
　　一、实训目的 ············ 236
　　二、实训设备 ············ 237
　　三、实训拓扑 ············ 237
　　四、实训要求 ············ 237
实验十三　OSPF 邻居认证配置 ······ 238
　　一、实验目的 ············ 238
　　二、相关知识 ············ 238
　　三、实验设备 ············ 238
　　四、实验拓扑 ············ 238

五、实验要求·····················239
六、实验步骤·····················239
七、注意事项和排错···········241
八、配置序列·····················241
九、思考题························241
十、课后练习·····················241

实验十四 OSPF 路由汇总配置·····241
一、实验目的·····················241
二、相关知识·····················242
三、实验设备·····················242
四、实验拓扑·····················242
五、实验要求·····················242
六、实验步骤·····················242
七、注意事项和排错···········245
八、配置序列·····················245
九、思考题························246
十、课后练习·····················246

实验十五 NAT 地址转换配置······246
一、实验目的·····················246
二、相关知识·····················247
三、实验设备·····················247
四、实验拓扑·····················247
五、实验要求·····················247
六、实验步骤·····················247
七、注意事项和排错···········249
八、配置序列·····················249
九、思考题························250
十、课后练习·····················250

第 3 章 网络综合实践·····················251

实践一 二层接入三层连通加聚合实践项目·····················251
一、实践目的·····················251

二、实践拓扑·····················251
三、实践步骤·····················251

实践二 单臂路由加聚合实践项目·····252
一、实践目的·····················252
二、实践拓扑·····················252
三、实践步骤·····················252

实践三 动态路由 RIP-OSPF 实践项目·····················253
一、实践目的·····················253
二、实践拓扑·····················253
三、实践步骤·····················254

实践四 路由与多层交换机间的 RIP-OSPF-静态路由重点分布实践项目·····················254
一、实践目的·····················254
二、实践拓扑·····················254
三、实践步骤·····················254

实践五 路由与多层交换机间的静态路由-OSPF-RIP 实践项目·········257
一、实践目的·····················257
二、实践拓扑·····················257
三、实践步骤·····················258

实践六 特定网络综合设计与实现实践项目·····················259
一、实践目的·····················259
二、实践拓扑·····················259
三、实践步骤·····················260
四、实验要求·····················260

附录·····················261

一、交换机相关配置命令详解·····················261
二、路由器相关配置命令详解·····················280

第1章 交换机实验

实验一 交换机带外管理

一、实验目的

1. 熟悉普通二层交换机的外观。
2. 了解普通二层交换机各端口的名称和作用。
3. 了解交换机最基本的管理方式——带外管理的方法。

二、相关知识

网络设备的管理方式可以简单地分为带外(out-of-band)管理和带内(in-band)管理两种管理模式。所谓带内管理,是指网络的管理控制信息与用户网络的承载业务信息通过同一个逻辑信道传送,简而言之,就是占用业务带宽;而在带外管理模式中,网络的管理控制信息与用户网络的承载业务信息在不同的逻辑信道传送,也就是设备提供专门用于管理的带宽。

目前很多高端的交换机都带有带外网管接口,使网络管理的带宽和业务带宽完全隔离,互不影响,构成单独的网管网。

通过 Console 口管理是最常用的带外管理方式,通常用户会在首次配置交换机或者无法进行带内管理时使用带外管理方式。

带外管理方式也是使用频率最高的管理方式。带外管理时,可以采用Windows 操作系统自带的超级终端程序来连接交换机,当然,用户也可以采用自己熟悉的终端程序。

Console 口:也叫配置口,用于接入交换机内部对交换机进行配置。

Console 线:交换机包装箱中的标配线缆,用于连接 Console 口和配置终端。

三、实验设备

1. DCS-3926S 交换机 1 台。
2. PC 机 1 台。
3. 交换机 Console 线 1 根。

四、实验拓扑

该实验拓扑结构如图 1-1 所示。

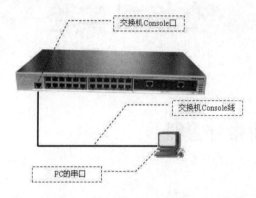

图 1-1　将 PC 机的串口和交换机的 Console 口用 Console 线连接

五、实验要求

1. 正确认识交换机上各端口名称。
2. 熟练掌握使用交换机 Console 线连接交换机的 Console 口和 PC 的串口。
3. 熟练掌握使用超级终端进入交换机的配置界面。

六、实验步骤

第一步：认识交换机的端口。

交换机的端口外观如图 1-2 所示，"0/0/1"中的第一个"0"表示堆叠中的第一台交换机，如果是"1"，就表示第 2 台交换机；第 2 个"0"表示交换机上的第 1 个模块(DCS-3926S 交换机有 3 个模块：网络端口模块 0(M0)，模块 1(M1)，模块 2(M2)；最后的"1"表示当前模块上的第 1 个网络端口)。

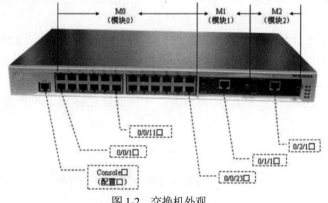

图 1-2　交换机外观

"0/0/1"表示用户使用的是堆叠中第 1 台交换机上的第 1 个网络端口模块上的第 1 个网络端口。默认情况下，如果不存在堆叠，交换机总会认为自己是第 0 台交换机。

第二步：连接 Console 线。

拔插 Console 线时注意保护交换机的 Console 口和 PC 的串口，不要带电拔插。

第三步：使用超级终端连入交换机。

(1) 启动 Windows 系统，选择"开始"|"程序"|"附件"|"通讯"|"超级终端"命令，打开"连接描述"对话框，如图 1-3 所示。在"名称"文本框中输入"DCS-3926S"，设置新建连接的名称(系统会将这个连接保存在附件的通信栏中，以便用户下次使用)，然后单击"确定"按钮。

图 1-3　"连接描述"对话框

(2) 在打开的"连接到"对话框中选择所使用的端口号。对话框第一行的"DCS-3926S"是在"连接描述"对话框中设置的"名称"，最后一行"连接时使用"下拉列表框默认设置连接在"COM1"口上(其他的选项视用户实际连接的端口而定)，如图 1-4 所示。

图 1-4　"连接到"对话框

(3) 在图 1-4 中单击"确定"按钮，打开"COM1 属性"对话框设置端口属性，如图 1-5 所示。单击对话框右下方的"还原为默认值"按钮，将"每秒位数"设置为 9600，"数据位"设置为 8，"奇偶校验"设置为无，"停止位"设置为 1，"数据流控制"设置为"无"，然后单击"确定"按钮。

图 1-5　设置 COM1 属性

(4) 此时，如果 PC 串口与交换机的 Console 口连接正确，在打开的超级终端中按回车键，将显示如图 1-6 所示的界面，表示已经进入了交换机，用户可以对交换机输入指令进行查看。

图 1-6　交换机配置界面

(5) 用户成功进入交换机的配置界面后，可以对交换机进行必要的配置。使用 show version 命令可以查看交换机的软硬件版本信息，如图 1-7 所示。

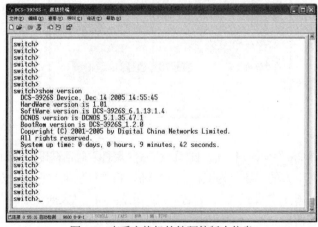

图 1-7　查看交换机的软硬件版本信息

(6) 使用 show running 命令查看当前配置。

```
Switch>enable                    ！进入特权配置模式(详见本章实验二)
switch#show running-config
Current configuration:
!
    hostname switch
!
Vlan 1
!
Interface Ethernet0/0/1
!
Interface Ethernet0/0/2
!
……
Interface Ethernet0/0/23
!
Interface Ethernet0/0/24
!
!
switch#
```

七、课后练习

1. 尝试用笔记本电脑对交换机进行带外管理。
2. 熟悉常用 show 命令。
(1) show version：显示交换机版本信息。
(2) show flash：显示保存在 flash 中的文件及大小。
(3) show arp：显示 ARP 映射表。
(4) show history：显示用户最近输入的历史命令。
(5) show rom：显示启动文件及大小。
(6) show running-config：显示当前运行状态下生效的交换机参数配置。
(7) show startup-config：显示当前运行状态下写在 Flash Memory 中的交换机参数配置，通常也是交换机下次启动时所用的配置文件。
(8) show switchport interface：显示交换机端口的 VLAN 端口模式和所属 VLAN 号及交换机端口信息。
(9) show interface ethernet 0/0/1：显示指定交换机端口的信息。

实验二 交换机的配置模式

一、实验目的

1. 了解交换机不同配置模式的功能。
2. 了解交换机不同配置模式的进入和退出方法。

二、相关知识

交换机的配置界面称为 CLI 界面。CLI 界面又称为命令行界面，和图形界面(GUI)相对应。CLI 的全称是 Command Line Interface，它由 Shell 程序提供，由一系列的配置命令组成，根据这些命令在配置管理交换机时所起的作用不同，Shell 将这些命令分类，不同类别的命令对应着不同的配置模式。

命令行界面是交换机调试界面中的主流界面，基本上所有的网络设备都支持命令行界面。国内外主流的网络设备供应商使用很相近的命令行界面，方便用户调试不同厂商的设备。神州数码网络产品的调试界面兼容国内外主流厂商的界面，和思科命令行接近，便于用户学习(只有少部分厂商使用自己独有的配置命令)。

交换机的配置模式有图 1-8 所示的几种。

图 1-8 交换机的配置模式

1. Setup 配置模式

当交换机首次出厂启动时，通常会自动进入 Setup 配置模式。该模式通常以菜单的形式呈现，用户可以在其中进行一些基本的配置，例如修改交换机提示符、配置 IP 地址以及启动 Web 服务等。

在大多数情况下，为了适应更复杂的网络环境配置需求，用户往往会跳过 Setup 模式，直接进入命令行界面进行详细配置。当用户退出 Setup 配置模式后，交换机将进入 CLI 配置界面。

在 Setup 模式中进行的配置在 CLI 配置界面中同样可以实现。然而，并非所有的交换机都具备 Setup 配置模式这一特性。

2. 一般用户配置模式

用户进入 CLI 界面后，首先进入的是一般用户配置模式，提示符通常为"switch>"，符号 ">" 为一般用户配置模式的提示符。当用户从特权用户配置模式使用 exit 命令退出时，可以回到一般用户配置模式。

在一般用户配置模式下，用户的权限受到限制，不能对交换机进行配置更改，只能执行一些基本的查询操作，例如查看交换机的系统时钟和版本信息。

大多数交换机都支持一般用户配置模式。

3. 特权用户配置模式

在一般用户配置模式下，使用 enable 命令(如果已经配置了进入特权用户配置模式的密码，则需输入相应的密码)，即可进入特权用户配置模式，提示符变为"switch#"。当用户从全局配置模式使用 exit 命令退出时，将返回到一般用户配置模式。此外，交换机还提供了 Ctrl+Z 快捷键，可以在任何配置模式下(一般用户配置模式除外)快速返回到特权用户配置模式。

大多数交换机都支持特权用户配置模式。

4. 全局配置模式

在进入特权用户配置模式后，只需使用 config 命令即可进入全局配置模式，提示符将变为"switch (config) #"。如果用户处于其他配置模式(如接口配置模式或 VLAN 配置模式)，可以使用 exit 命令返回到全局配置模式。

5. 接口配置模式

在全局配置模式下，通过输入 interface 命令即可进入相应的接口配置模式。交换机操作系统支持以下两种端口类型。

(1) CPU 端口：在二层交换机中，创建的 VLAN 接口通常用于管理，称为管理 VLAN。该接口将使用交换机的 MAC 地址。

(2) 以太网端口：物理以太网端口，用于数据传输。

因此，存在两种接口配置模式，如表 1-1 所示。

表 1-1 接口配置模式

接口类型	进入方式	提示符	可执行操作	退出方式
CPU 端口	在全局配置模式下，输入命令 interface vlan 1	switch(config-if-vlan)#	配置交换机的 IP 地址，设置管理 VLAN	使用 exit 命令即可返回全局配置模式
以太网端口	在全局配置模式下，输入命令 interface ethernet <interface-list>	switch(config-if)#	配置交换机提供的以太网接口的双工模式、速率、广播抑制等	使用 exit 命令即可返回全局配置模式

6. VLAN 配置模式

在全局配置模式下，使用命令 VLAN <vlan-id>即可进入对应的 VLAN 配置模式。在"CS-3926S"设备中输入所需创建的 VLAN 编号，即可进入该 VLAN 的配置模式。在此模式下，用户可以为本 VLAN 配置成员端口，并调整其各种属性。

三、实验设备

1. DCS-3926S 交换机 1 台。
2. PC 机 1 台。
3. Console 线 1 根。

四、实验拓扑

该实验拓扑结构如图 1-9 所示。

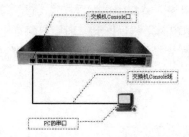

图 1-9　交换机与主机通过 Console 口连接图

五、实验要求

1. 熟悉 Setup 配置模式。
2. 熟悉一般用户配置模式。
3. 熟悉特权用户配置模式。
4. 了解全局配置模式。
5. 了解接口配置模式。
6. 了解 VLAN 配置模式。

六、实验步骤

第一步：Setup 模式的配置方法。

交换机出厂第一次启动时将进入"Setup Configuration"状态，如图 1-10 所示。在该状态下，用户可以选择进入 Setup 模式或者跳过 Setup 模式(输入"y"，再按回车键可以进入 Setup 模式)。

在进入主菜单之前，系统会提示用户选择配置界面的语言种类，对英文不熟悉的用户可以选择"1"，进入中文提示的配置界面，如图 1-11 所示(选择"0"将进入英文提示的配置界面)。

```
Please select language
[0]:English
[1]:中文
Selection(0|1)[0]:
```

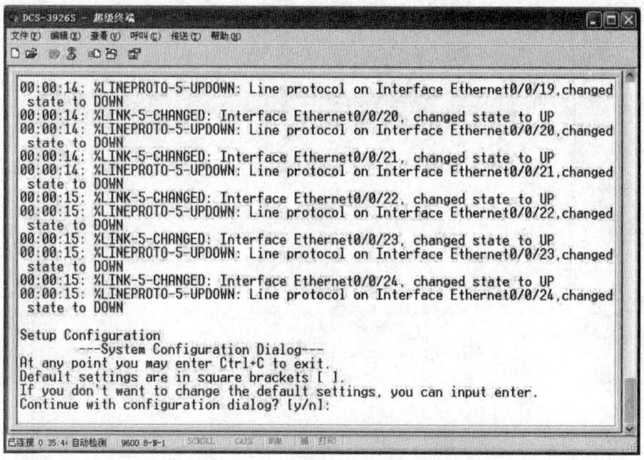

图 1-10 交换机启动界面

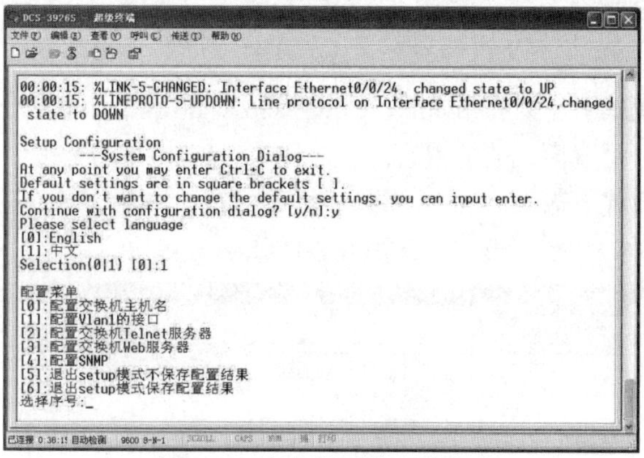

图 1-11 中文提示的配置界面

以下是 Setup 主菜单的提示：

Configure menu	！配置菜单
[0]:Config hostname	！配置交换机的名字
[1]:Config interface-Vlan1	！配置交换机的管理 IP
[2]:Config telnet-server	
[3]:Config web-server	
[4]:Config SNMP	
[5]:Exit setup configuration without saving	！不保存配置结果，退出 Setup 模式
[6]:Exit setup configuration after saving	！保存配置结果，退出 Setup 模式
Selection number:	

在图 1-11 所示的 Setup 主菜单上选择序号"5"，将在退出 Setup 模式的同时不保留 Setup 模式中的配置。

在图 1-11 所示的 Setup 主菜单上选择序号 "6"，将在退出 Setup 模式的同时保留用户在 Setup 模式下所做的配置。例如，用户在 Setup 模式下设置了 IP 地址、打开了 Web 服务，选择序号 "6" 退出 Setup 主菜单后，用户就可以通过 PC 对交换机进行 HTTP 管理配置。

第二步：一般用户配置模式的配置方法。

退出 Setup 模式即进入一般用户配置模式，一般用户配置模式也可以称为 ">" 模式。该模式的命令比较少，可以使用 "?" 命令来查看，如图 1-12 所示。

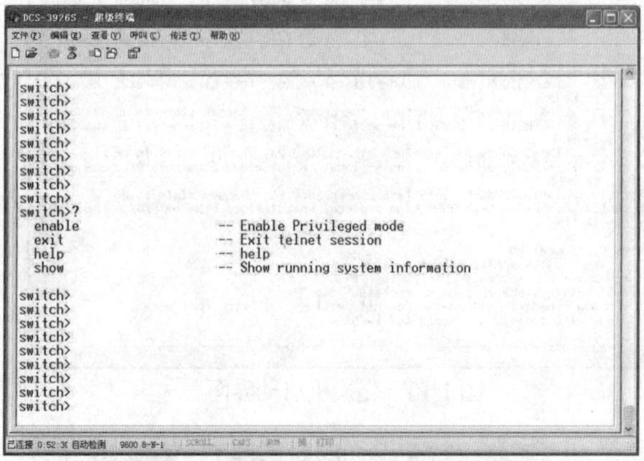

图 1-12　通过 "?" 命令使用帮助

在一般用户配置模式下，只有 enable、exit、help、show 这 4 个命令可以使用。

第三步：特权用户配置模式的配置方法。

在一般用户配置模式下输入 enable，进入特权用户配置模式。特权用户配置模式的提示符为 "#"，所以也称为 "#" 模式，该模式界面如图 1-13 所示。

图 1-13　特权用户配置模式

在特权用户配置模式下，用户可以查询交换机配置信息、各个端口的连接情况、收发数据统计等。而且进入特权用户配置模式后，可以进入全局模式对交换机的各项配置进行修改，因此必须设置进入特权用户配置模式的口令，防止非特权用户的非法使用。

第四步：全局配置模式的配置方法。

在特权模式下输入 config terminal、config t 或 config 就可以进入全局配置模式。全局配置模式也称为 config 模式。

```
switch#config terminal
switch(config)#
```

在全局配置模式下，用户可以对交换机进行全局性的配置，如对 MAC 地址表、端口镜像、创建 VLAN、启动 IGMP Snooping、GVRP、STP 等功能进行配置。用户在全局模式还可以通过命令进入端口对各个端口进行配置。

下面在全局配置模式下设置特定用户密码：

```
switch>enable
switch#config terminal                              ！进入全局配置模式，见第四步
switch(Config)#enable password level admin
Current password:                                   ！原密码为空，直接回车
New password:*****
Confirm new password:*****                          ！输入密码
switch(Config)#exit
switch#write
switch#
```

验证配置。
验证方法 1：重新进入交换机。

```
switch#exit                                         ！退出特权用户配置模式
switch>
switch>enable                                       ！进入特权用户配置模式
Password:*****
switch#
```

验证方法 2：使用 show 命令进行查看。

```
switch#show running-config
Current configuration:
enable password level admin 827ccb0eea8a706c4c34a16891f84e7b
  ！该行显示了已经为交换机配置了 enable 密码
    hostname switch
    Vlan 1
……                                                  ！省略部分显示
```

第五步：接口配置模式的配置方法。

```
switch(Config)#interface ethernet 0/0/1
switch(Config-Ethernet0/0/1)#                       ！已经进入以太端口 0/0/1 的接口
switch(Config)#interface vlan 1
switch(Config-If-Vlan1)#                            ！已经进入 VLAN 1 的接口，也就是 CPU 的接口
```

第六步：VLAN 配置模式的配置方法。

```
switch(Config)#vlan 100
switch(Config-Vlan100)#
```

验证配置：

```
switch(Config-Vlan100)#exit
switch(Config)#exit
switch#show vlan
VLAN    Name       Type     Media    Ports
1       default    Static   ENET     Ethernet0/0/1    Ethernet0/0/2
                                     Ethernet0/0/3    Ethernet0/0/4
                                     Ethernet0/0/5    Ethernet0/0/6
                                     Ethernet0/0/7    Ethernet0/0/8
                                     Ethernet0/0/9    Ethernet0/0/10
                                     Ethernet0/0/11   Ethernet0/0/12
                                     Ethernet0/0/13   Ethernet0/0/14
                                     Ethernet0/0/15   Ethernet0/0/16
                                     Ethernet0/0/17   Ethernet0/0/18
                                     Ethernet0/0/19   Ethernet0/0/20
                                     Ethernet0/0/21   Ethernet0/0/22
                                     Ethernet0/0/23   Ethernet0/0/24
100     VLAN0100   Static   ENET
switch#
！可以看到，已经新增了一个"VLAN 100"的信息
```

第七步：实验结束后，取消 enable 密码。

如果不取消 enable 密码，下一批的同学将没有办法做实验。因此，所有个人在实验中设定的密码都应该在实验完成之后取消，为后面实验的同学带来方便，这也是一个网络工程师基本的素质。

```
switch(Config)#no enable password level admin
Input password:*****
switch(Config)#
```

七、注意事项和排错

1. 特定的命令存在于特定的配置模式下。大家在进行配置时不仅需要输入正确的命令，还需要知道该命令是否在正确的配置模式下。

2. 当用户不知道该命令是否正确时，可以使用"?"来咨询交换机。

八、思考题

1. 为什么 enable 密码在 show 命令显示时，不是出现配置的密码，而是一大堆不认识的字符？

2. 当不能确定一个命令是否存在于某个配置模式下时，应该怎么查询？

九、课后练习

1. 进入各个配置模式并退出。

2. 设置特权用户配置模式的 enable 密码为"digitalchina"。
3. 在实验结束后取消 enable 密码。

实验三 交换机 CLI 界面调试技巧

一、实验目的

1. 熟悉交换机 CLI 界面。
2. 了解基本的命令格式。
3. 了解交换机 CLI 界面的部分调试技巧。

二、应用环境

所有其他的实验都需要使用到本实验所讲述的内容,熟悉本实验将会对其他实验的操作带来方便。

三、实验设备

1. DCS-3926S 交换机 1 台。
2. PC 机 1 台。
3. Console 线 1 根。

四、实验拓扑

该实验拓扑结构如图 1-14 所示。

图 1-14 实验拓扑图

五、实验要求

1. 熟悉帮助功能。
2. 了解交换机对输入的检查。
(1) 成功返回信息。

(2) 错误返回信息。
3. 熟练使用不完全匹配功能。
4. 熟悉以下常用配置技巧。
(1) 命令简写。
(2) 命令完成。
(3) 命令查询。
(4) 否定命令的作用。
(5) 命令历史。

六、实验步骤

第一步：使用"?"。

switch#show v?	！查看 v 开头的命令
version vlan	！只有两条命令：show version 和 show vlan
switch#show version	！查看交换机版本信息

第二步：查看错误信息。

switch#show v	！直接输入 show v，回车
> Ambiguous command!	
switch#	！根据已有输入可以产生至少两种不同的解释
switch#show valn	！show vlan 写成了 show valn
> Unrecognized command or illegal parameter!	！不识别的命令
switch#	

第三步：不完全匹配。

switch#show ver	！应该是 show version，没有输全，但是无歧义即可
DCS-3926S Device, Aug 23 2005 09:35:31	
HardWare version is 1.01	
SoftWare version is DCS-3926S_6.1.12.0	
DCNOS version is DCNOS_5.1.35.42	
BootRom version is DCS-3926S_1.2.0	
Copyright (C) 2001-2005 by Digital China Networks Limited.	
All rights reserved.	
System up time: 0 days, 0 hours, 22 minutes, 43 seconds.	
switch#	

第四步：Tab 的用途。

switch#show v	！show v，按 Tab 键，出错，因为有 show vlan，有歧义
> Ambiguous command!	
switch#show ver	！show ver，按 Tab 键补全命令
DCS-3926S Device, Aug 23 2005 09:35:31	
HardWare version is 1.01	
SoftWare version is DCS-3926S_6.1.12.0	
DCNOS version is DCNOS_5.1.35.42	

```
BootRom version is DCS-3926S_1.2.0
Copyright (C) 2001-2005 by Digital China Networks Limited.
All rights reserved.
System up time: 0 days, 0 hours, 35 minutes, 56 seconds.
switch#
```

只有当前命令正确的情况下，才可以使用 Tab 键。也就是说，如果命令未输入完整，而 Tab 键却没有起作用，说明当前命令中可能存在错误或参数错误等，此时需要仔细排查。

第五步：否定命令"no"。

```
switch#config                                    ！进入全局配置模式
switch(Config)#vlan 10                           ！创建 VLAN 10 并进入 VLAN 配置模式
switch(Config-Vlan10)#exit                       ！退出 VLAN 配置模式
switch(Config)#show vlan                         ！查看 VLAN
> Unrecognized command or illegal parameter!     ！该命令不在全局配置模式下
switch(Config)#exit                              ！退出全局配置模式
switch#show vlan                                 ！查看 VLAN 信息
VLAN    Name         Type       Media    Ports
1       default      Static     ENET     Ethernet0/0/1      Ethernet0/0/2
                                         Ethernet0/0/3      Ethernet0/0/4
                                         Ethernet0/0/5      Ethernet0/0/6
                                         Ethernet0/0
                                         Ethernet0/0/9      Ethernet0/0/10
                                         Ethernet0/0/11     Ethernet0/0/12
                                         Ethernet0/0/13     Ethernet0/0/14
                                         Ethernet0/0/15     Ethernet0/0/16
                                         Ethernet0/0/17     Ethernet0/0/18
                                         Ethernet0/0/19     Ethernet0/0/20
                                         Ethernet0/0/21     Ethernet0/0/22
                                         Ethernet0/0/23     Ethernet0/0/24
10      VLAN0010     Static     ENET     ！有 VLAN 10 的存在
switch#config
switch(Config)#no vlan 10                        ！使用 no 命令删掉 VLAN 10
switch(Config)#exit
switch#show vlan
VLAN    Name         Type       Media    Ports
----    ----------   --------   ------   ---------------------------------
1       default      Static     ENET     Ethernet0/0/1      Ethernet0/0/2
                                         Ethernet0/0/3      Ethernet0/0/4
                                         Ethernet0/0/5      Ethernet0/0/6
                                         Ethernet0/0/7      Ethernet0/0/8
                                         Ethernet0/0/9      Ethernet0/0/10
                                         Ethernet0/0/11     Ethernet0/0/12
                                         Ethernet0/0/13     Ethernet0/0/14
                                         Ethernet0/0/15     Ethernet0/0/16
                                         Ethernet0/0/17     Ethernet0/0/18
                                         Ethernet0/0/19     Ethernet0/0/20
                                         Ethernet0/0/21     Ethernet0/0/22
```

	Ethernet0/0/23	Ethernet0/0/24
switch#	！VLAN 10 不见了，已经删掉了	

交换机中大部分命令的逆命令都是采用 no 命令的模式，还有一种否定模式是 enable 和 disable 的相反。

第六步：使用上下箭头键(↑↓)可以快速选择之前输过的命令，从而节省时间。

七、课后练习

本实验的内容需要课后大量练习 VLAN 的划分及端口的验证联通。

实验四　交换机恢复出厂设置及其基本配置

一、实验目的

1. 了解交换机的文件管理。
2. 了解什么时候需要将交换机恢复成出厂设置。
3. 了解交换机恢复出厂设置的方法。
4. 了解交换机的一些基本配置命令。

二、相关知识

交换机恢复出厂设置通常意味着将设备的所有配置重置为出厂时的状态，通常包括删除所有用户自定义的配置，如 VLAN 配置、端口配置、密码设置等。交换机恢复出厂设置的实际应用环境如下。

(1) 教学楼的 DCS3926S 交换机坏了，网络管理员把实验楼的一台交换机拿过去先用着。这台交换机的配置是按照实验楼的环境设置的，我需要改成教学楼的环境，一条条修改比较麻烦，也不能保证正确，可以清空交换机的所有配置，恢复到刚刚出厂的状态。

(2) 我正在配置一台 DCS-3926S 交换机，做了很多功能的配置，完成之后发现它不能正常工作。问题出在哪里了？我检查了很多遍都没有发现错误。排错的难度远远大于重新做配置，不如清空交换机的所有配置，恢复到刚刚出厂的状态。

(3) 上一节网络实验课的同学们刚刚做完实验，已经离去。桌上的交换机已经配置过，为了不影响这节课的实验，可以将交换机恢复到出厂设置，然后按照自己的思路重新配置，从而验证自己的配置思路是否能够生效。

三、实验设备

1. DCS-3926S 交换机 1 台。
2. PC 机 1 台。
3. Console 线 1 根。

四、实验拓扑

该实验拓扑结构如图 1-15 所示。

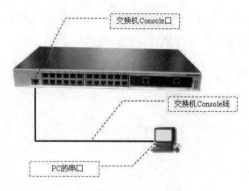

图 1-15 实验拓扑图

五、实验要求

1. 先为交换机设置 enable 密码，并确认密码设置成功。
2. 对交换机进行恢复出厂设置，重新启动后发现 enable 密码已消失，说明恢复成功。
3. 了解 show flash 命令以及显示内容。
4. 了解 clock set 命令以及显示内容。
5. 了解 hostname 命令以及显示内容。
6. 了解 language 命令以及显示内容。

六、实验步骤

第一步：在交换机上设置 enable 密码(详见实验二)。

```
switch>enable
switch#config t                              ! 进入全局配置模式
switch(Config)#enable password level admin
Current password:                            ! 原密码为空，直接回车
New password:*****
Confirm new password:*****
switch(Config)#exit
switch#write
switch#
```

验证配置。
验证方法 1：重新进入交换机。

```
                                             ! 输入密码
switch#exit                                  ! 退出特权用户配置模式
```

```
switch>
switch>enable                              !进入特权用户配置模式
Password:*****
switch#
```

验证方法 2：使用 show 命令进行查看。

```
switch#show running-config
Current configuration:
!
    enable password level admin 827ccb0eea8a706c4c34a16891f84e7b    !已经为交换机配置了 enable 密码
    hostname switch
!
!
Vlan 1
    vlan 1
!
!
……                                                !省略部分显示
```

第二步：清空交换机的配置。

```
switch>enable                              !进入特权用户配置模式
switch#set default                         !使用 set default 命令
Are you sure? [Y/N] = y                    !是否确认？
switch#write                               !清空 startup-config 文件
switch #show startup-config                !显示当前的 startup-config 文件
This is first time start up system.        !系统提示此启动文件为出厂默认配置
switch#reload                              !重新启动交换机
Process with reboot? [Y/N] y
```

验证测试。

验证方法 1：重新进入交换机。

```
switch>
switch>enable
switch#                                    !已经不需要输入密码就可进入特权模式
```

验证方法 2：使用 show 命令进行查看。

```
switch#show running-config
Current configuration:
!
    hostname switch                        !已经没有 enable 密码显示了
!
```

```
Vlan 1
    vlan 1
!
……                                          ! 省略部分显示
```

第三步：使用 show flash 命令。

```
switch#show flash
file name           file length
nos.img             1720035 bytes        ! 交换机软件系统
startup-config      0 bytes              ! 启动配置文件
running-config      783 bytes            ! 当前配置文件
switch#
```

第四步：设置交换机系统日期和时钟。

```
switch#clock set ?                              ! 使用？查询命令格式
   <HH:MM:SS>                  -- Time
switch#clock set 15:29:50                       ! 配置当前时间
Current time is MON JAN 01 15:29:50 2001        ! 配置完即有显示，注意年份不对
switch#clock set 15:29:50 ?                     ! 使用"？"查询，原来命令没有结束
   <YYYY.MM.DD>                -- Date <year:2000-2035>
   <CR>
switch#clock set 15:29:50 2006.01.16            ! 配置当前年月日
Current time is MON JAN 16 15:29:50 2006        ! 正确显示
```

验证配置：

```
switch#show clock                               ! 再用 show 命令验证
Current time is MON JAN 16 15:29:55 2006
switch#
```

第五步：设置交换机命令行界面的提示符(设置交换机的名称)。

```
switch#
switch#config
switch(Config)#hostname DCS-3926S-BD1           ! 配置名称
DCS-3926S-BD1(Config)#exit                      ! 无需验证，即配即生效
DCS-3926S-BD1#
DCS-3926S-BD1#
```

第六步：配置显示的帮助信息的语言类型。

```
DCS-3926S-BD1#language ?
   chinese                     -- Chinese
   english                     -- English
DCS-3926S-BD1#language chinese
```

```
DCS-3926S-BD1#language ?                    ！请注意再使用 "?" 时，帮助信息已经成了中文
    chinese              -- 汉语
    english              -- 英语
```

七、注意事项和排错

1. 在执行 set default 命令恢复出厂设置后，务必使用 write 命令保存更改。重新启动设备后，设置才会生效。

2. 这几个命令中，hostname 命令是在全局配置模式下进行配置的。

八、配置序列

```
DCS-3926S-BD1#show running-config
Current configuration:
!
    hostname DCS-3926S-BD1         ！上述配置中，只有 hostname 命令的设置会在 show run 命令
                                   的输出中显示
!
!
Vlan 1
    vlan 1
!
Interface Ethernet0/0/1
……                                ！省略部分
Interface Ethernet0/0/23
!
Interface Ethernet0/0/24
!
DCS-3926S-BD1#
```

九、思考题

1. 为什么在第三步中 show flash 的显示中，startup-config 文件的大小为 0 bytes？

2. 如何才能使 startup-config 文件与 running-config 文件保持一致？

十、课后练习

1. 将交换机的 enable 密码设置为 digitalchina。

2. 将交换机的时钟设置为当前时间。

3. 将交换机的名称设置为 digitalchina-3926S。

4. 将交换机的帮助信息设置为中文。

5. 将交换机恢复为出厂设置。

实验五 使用 Telnet 方式管理交换机

一、实验目的

1. 了解什么是带内管理；
2. 熟练掌握如何使用 Telnet 方式管理交换机。

二、相关知识

使用 Telnet 协议管理交换机是一种基于网络的远程访问和管理方式。Telnet 管理是一种带内管理方式，它允许通过网络连接的设备对交换机进行配置和管理。当交换机的配置发生变化而导致带内管理失效时，必须转用带外管理来进行配置管理。

这种管理模式适用于以下场景：学校有 20 台交换机支撑着校园网的运营，这 20 台交换机分别放置在学校的不同位置。网络管理员需要对这 20 台交换机进行管理，通过前面学习的知识，我们可以通过带外管理的方式也就是通过 Console 口去管理，那么管理员需要携带自己的笔记本电脑，并带着 Console 线去学校的不同位置去调试每台交换机，十分麻烦。校园网既然是互联互通的，在网络的任何一个信息点都应该能访问其他的信息点，管理员可以通过 Telnet 方式，在办公室内远程调试全校所有的交换机。

三、实验设备

1. DCS-3926S 交换机 1 台。
2. PC 机 1 台。
3. Console 线 1 根。
4. 直通网线 1 根。

四、实验拓扑

该实验拓扑结构如图 1-16 所示。

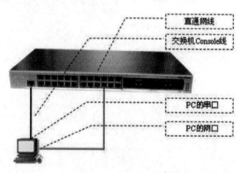

图 1-16 实验拓扑图

五、实验要求

1. 按照拓扑图连接网络。
2. PC 和交换机的 24 口用网线相连。
3. 交换机的管理 IP 为 192.168.2.100/24。
4. PC 网卡的 IP 地址为 192.168.2.101/24。

六、实验步骤

第一步：交换机恢复出厂设置，设置正确的时钟和标识符(详见实验四)。

```
switch#set default
Are you sure? [Y/N] = y
switch#write
switch#reload
Process with reboot? [Y/N] y
switch#clock set 15:29:50 2006.01.16
Current time is MON JAN 16 15:29:50 2006
switch#
switch#config
switch(Config)#hostname DCS-3926S
DCS-3926S(Config)#exit
DCS-3926S#
```

第二步：给交换机设置 IP 地址即管理 IP。

```
DCS-3926S#config
DCS-3926S(Config)#interface vlan 1                          ！进入 VLAN 1 接口
02:20:17: %LINK-5-CHANGED: Interface Vlan1, changed state to UP
DCS-3926S(Config-If-Vlan1)#ip address 192.168.2.100 255.255.255.0   ！配置地址
DCS-3926S(Config-If-Vlan1)#no shutdown                      ！激活 VLAN 接口
DCS-3926S(Config-If-Vlan1)#exit
DCS-3926S(Config)#exit
DCS-3926S#
```

验证配置：

```
DCS-3926S#show run
Current configuration:
!
   hostname DCS-3926S
!
Vlan 1
```

```
    vlan 1
!
Interface Ethernet0/0/1
……
Interface Ethernet0/0/24
!
interface Vlan1
    interface vlan 1
    ip address 192.168.2.100 255.255.255.0        !已经配置好交换机 IP 地址
!
DCS-3926S#
```

第三步：为交换机设置授权 Telnet 用户。

```
DCS-3926S#config
DCS-3926S(Config)#telnet-user xuxp password 0 digital
DCS-3926S(Config)#exit
DCS-3926S#
```

验证配置：

```
DCS-3926S#show run
Current configuration:
!
    hostname DCS-3926S
!
    telnet-user xuxp password 0 digital
!
Vlan 1
    vlan 1
!
Interface Ethernet0/0/1
……
Interface Ethernet0/0/24
!
interface Vlan1
    interface vlan 1
    ip address 192.168.2.100 255.255.255.0
!
DCS-3926S#
```

第四步：配置主机的 IP 地址，在本实验中要与交换机的 IP 地址在一个网段。
主机 IP 地址配置如图 1-17 所示。

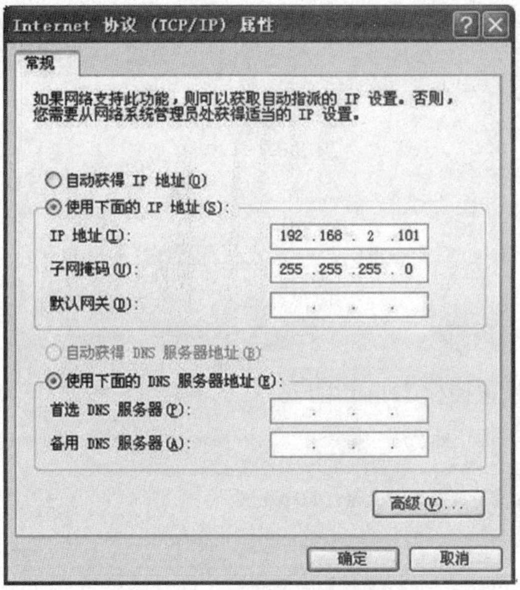

图 1-17 主机 IP 地址配置

验证配置：可以在 PC 主机的 DOS 命令行中使用 ipconfig 命令查看 IP 地址配置，结果如图 1-18 所示。

图 1-18 使用 ipconfig 命令查看 IP 地址

第五步：验证主机与交换机是否连通。
验证方法 1：在交换机中 ping 主机。

```
DCS-3926S#ping 192.168.2.101

Type ^c to abort.

Sending 5 56-byte ICMP Echos to 192.168.2.101, timeout is 2 seconds.
!!!!!
Success rate is 100 percent (5/5), round-trip min/avg/max = 1/1/1 ms
DCS-3926S#
```

稍后，将出现 5 个 "！"，表示已经连通。

第 1 章 交换机实验

验证方法 2：在 PC 主机 DOS 命令行中 ping 交换机，若显示如图 1-19 所示的结果则表示连通。

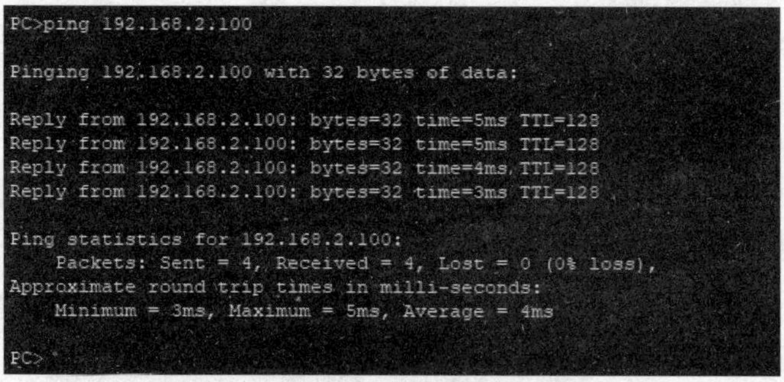

图 1-19 使用 ping 命令查看网络连通性

第六步：使用 Telnet 登录。

启动 Windows 系统，选择"开始" | "运行"命令，运行 Windows 自带的 Telnet 客户端程序，输入 Telnet 的目的地址后单击"确定"按钮，如图 1-20 所示。

图 1-20 Windows 自带的 Telnet 客户端程序

在打开的窗口中输入正确的登录名和密码(登录名为 xuxp，密码为 digital)后按回车键。Telnet 登录界面如图 1-21 所示。

图 1-21 Telnet 登录界面

对交换机进行进一步配置后，本实验完成。

七、注意事项和排错

1. 默认情况下，交换机所有端口都属于 VLAN 1。我们通常把 VLAN 1 作为交换机的管理 VLAN，因此 VLAN 1 接口的 IP 地址就是交换机的管理地址。

2. 密码只能是 1~8 个字符。

3. 若要删除一个 Telnet 用户，可以在 config 模式下使用 no telnet-user 命令。

八、配置序列

```
DCS-3926S#show running-config
Current configuration:
!
    hostname DCS-3926S
!
    telnet-user xuxp password 0 digital
!
Vlan 1
    vlan 1
!
Interface Ethernet0/0/1
......
Interface Ethernet0/0/24
interface Vlan1
    interface vlan 1
    ip address 192.168.2.100 255.255.255.0
!
DCS-3926S#
```

九、思考题

1. 二层交换机可以配置多少个 IP 地址？为什么？
2. 是否可以为 VLAN 2 配置 IP 地址？
3. 在 telnet-user xuxp password 0 digital 命令中，将"0"换成"7"会产生什么效果？

十、课后练习

1. 删除 xuxp 用户(不允许使用 set default 命令)。
2. 设置交换机的管理 IP 为 10.1.1.1，子网掩码为 255.255.255.0。
3. 使用用户名 aaa 和密码 bbb，并选择"7"作为 Telnet 密码的加密方式进行配置。

实验六 使用 Web 方式管理交换机

一、实验目的

1. 熟练掌握如何为交换机设置 Web 方式管理。
2. 熟练掌握如何进入交换机 Web 管理方式。

3. 了解交换机的 Web 配置界面，并能够进行部分操作。

二、相关知识

　　Web 方式管理交换机是一种基于图形用户界面的远程管理方法。用户可以通过 Web 浏览器访问交换机的 IP 地址，输入用户名和密码后进入交换机的 Web 管理界面进行操作。通常，Web 管理会通过 HTTPS 协议加密数据传输，以确保安全性。

　　Web 方式管理交换机主要适用于以下几个场景。

　　(1) 当交换机位于远程位置，管理员无法直接访问时，Web 管理提供了一种便捷的远程配置和管理手段。

　　(2) 对于不熟悉命令行界面的用户，Web 界面提供了一种更直观、更易于操作的配置方式。

　　(3) Web 界面通常提供实时监控功能，如流量统计和端口状态，便于管理员进行故障排除。

三、实验设备

1. DCS-3926S 交换机 1 台。
2. PC 机 1 台。
3. Console 线 1 根。
4. 直通网线 1 根。

四、实验拓扑

　　该实验拓扑结构如图 1-22 所示。

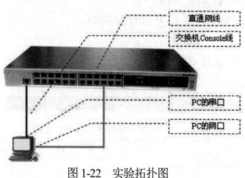

图 1-22　实验拓扑图

五、实验要求

1. 按照拓扑图连接网络。
2. PC 和交换机的 24 口用网线相连。
3. 交换机的管理 IP 为 192.168.2.100/24。
4. PC 网卡的 IP 地址为 192.168.2.101/24。

六、实验步骤

第一步：将交换机恢复出厂设置，设置正确的时钟和标识符(详见实验四)。
第二步：为交换机配置管理 IP(详见实验五)。

```
DCS-3926S#config
DCS-3926S(Config)#interface vlan 1
DCS-3926S(Config-If-Vlan1)#ip address 192.168.2.100 255.255.255.0
DCS-3926S(Config-If-Vlan1)#no shutdown
DCS-3926S(Config-If-Vlan1)#exit
DCS-3926S(Config)#
```

第三步：启动交换机 Web 服务。

```
DCS-3926S#config
DCS-3926S(Config)#ip http server
web server is on                    !表明已经成功启动
DCS-3926S(Config)#
```

第四步：设置交换机授权 HTTP 用户。

```
DCS-3926S(Config)#web-user admin password 0 digital
DCS-3926S(Config)#
```

第五步：配置主机的 IP 地址，在本实验中要与交换机的 IP 地址在一个网段(详见实验五)。配置主机的地址为：192.168.2.101。
第六步：验证主机与交换机是否连通(详见实验五)。
验证方法 1：在交换机中 ping 主机。
验证方法 2：在主机 DOS 命令行中 ping 交换机。
第七步：使用 http 登录。

启动 Windows 系统，选择"开始"|"运行"命令，在打开的"运行"对话框中输入目标地址，如图 1-23 所示。

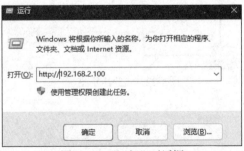

图 1-23 "运行"对话框

在打开的窗口中输入正确的登录名和密码(登录名为 admin，密码为 digital)，然后单击"登录"按钮，如图 1-24 所示。

第 1 章 交换机实验

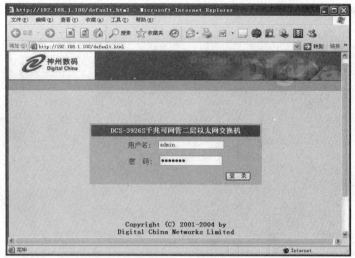

图 1-24 登录界面

此时，将打开图 1-25 所示的交换机 Web 调试界面的主界面。

图 1-25 交换机 Web 调试界面的主界面

对交换机进行进一步配置后，本实验完成。

七、注意事项和排错

1. 使用 Telnet 和 Web 方式调试有以下两个相同的前提条件。
(1) 交换机开启该功能并进行用户设置。
(2) 交换机和主机之间需要实现互联互通，确保能够通过 ping 命令进行测试。
2. 有时候交换机和主机的地址配置都正确，但仍然无法 ping 通。这时，如果排除了硬件问题，可能的原因是主机的 Windows 操作系统启用了防火墙。可以尝试关闭防火墙来解决问题。

八、配置序列

```
DCS-3926S#show run
Current configuration:
!
    hostname DCS-3926S
!
    telnet-user xuxp password 0 digital
!
Vlan 1
    vlan 1
!
Interface Ethernet0/0/1
......
Interface Ethernet0/0/24
!
interface Vlan1
    interface vlan 1
    ip address 192.168.2.100 255.255.255.0
!
    ip http server
    web-user admin password 0 digital
!
DCS-3926S#
```

九、思考题

1. 如何关闭 Web 服务？
2. 如何删除 Web 用户？

十、课后练习

1. 重新配置一个 Web 用户。
2. 在 Web 界面下，找出前几个实验中配置命令的位置，并进行相应的修改。

实训项目一　交换机配置模式及管理

一、实训目的

1. 熟悉网络实训环境及网络仿真平台。
2. 掌握交换机的六种配置模式的操作方法和技能。
3. 掌握交换机名称设置、密码管理和重置等操作方法。

二、实训设备

1. DCS-3926S 交换机 1 台。
2. PC 机 1 台。
3. 直通网线 1 根。

三、实训拓扑

该实验拓扑结构如图 1-26 所示。

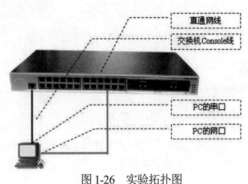

图 1-26 实验拓扑图

四、实训内容

1. 实验台和网络实训环境。
2. 交换机 Setup 配置模式配置。
3. 交换机一般用户配置模式配置。
4. 交换机特权用户配置模式配置。
5. 交换机全局配置模式配置。
6. 交换机接口配置模式配置。
7. 交换机 VLAN 配置模式配置。
8. 交换机重置、名称修改和密码配置。

五、实训步骤

第一步：在交换机的 Setup 配置模式下进行配置。

通常情况下，Setup 配置模式在交换机首次启动时进入，或者通过重启交换机并在启动过程中按下特定键进入。选择"y"进入 Setup 模式，选择"n"则退出 Setup 模式。

第二步：在交换机的一般用户配置模式下进行配置。

```
switch>                              ! 一般模式
```

第三步：在交换机的特权用户配置模式下进行配置。

```
switch> enable                       ! 在一般模式下输入 enable 进入特权模式
```

switch#	！特权模式

第四步：在交换机的全局配置模式下进行配置。

switch#config	！在特权模式下输入 config 进入全局配置模式
switch(config)#	

第五步：在交换机的接口配置模式下进行配置。

switch(config)#interface vlan 1	！在全局模式下输入 interface vlan 1 进入 CPU 接口配置模式
switch(Config-If-Vlan1)#	

第六步：在交换机的 VLAN 配置模式下进行配置。

switch(Config-If-Vlan1)#exit	！退出本模式
switch(config)#interface ethernet 0/0/15	！在全局模式下输入 interface ethernet 0/0/15 进入交换机接口配置模式
switch(Config-Ethernet0/0/15)#exit	！输入 exit 退出本模式
switch(config)#vlan 1	！在全局模式下输入 vlan 1 进入交换机 VLAN 配置模式
switch(Config-Vlan1)#exi	！退出本模式

第七步：重置交换机配置，并修改用户名和密码。

switch>enable	！进入特权用户配置模式
switch#set default	！使用 set default 命令
Are you sure? [Y/N] = y	！是否确认？
switch#write	！清空 startup-config 文件
switch #show startup-config	！显示当前的 startup-config 文件
This is first time start up system.	！系统提示此启动文件为出厂默认配置
switch#reload	！重新启动交换机
Process with reboot? [Y/N] y	
switch>enable	
switch#config	！进入全局配置模式
switch(Config)#hostname S1	
S1(Config)#enable password level admin	
Current password:	！原密码为空，直接按回车键
New password:*****	
Confirm new password:*****	
S1(Config)#exit	
S1#write	
S1#	

思科设备的配置命令如下：

Switch>
Switch>enable

```
Switch#configure terminal
Switch(config)#
Switch(config)#exit
Switch#
Switch#?
Exec commands:
  clear        Reset functions
  clock        Manage the system clock
  configure    Enter configuration mode
  connect      Open a terminal connection
  copy         Copy from one file to another
  debug        Debugging functions (see also 'undebug')
  delete       Delete a file
  dir          List files on a filesystem
  disable      Turn off privileged commands
  disconnect   Disconnect an existing network connection
  enable       Turn on privileged commands
  erase        Erase a filesystem
  exit         Exit from the EXEC
  logout       Exit from the EXEC
  more         Display the contents of a file
  no           Disable debugging informations
  ping         Send echo messages
  reload       Halt and perform a cold restart
  resume       Resume an active network connection
  setup        Run the SETUP command facility
  show         Show running system information
  ssh          Open a secure shell client connection
  telnet       Open a telnet connection
  terminal     Set terminal line parameters
  traceroute   Trace route to destination
  undebug      Disable debugging functions (see also 'debug')
  vlan         Configure VLAN parameters
  write        Write running configuration to memory, network, or terminal
Switch#show running-config
Building configuration...
Current configuration : 977 bytes
!
version 12.1
no service timestamps log datetime msec
no service timestamps debug datetime msec
no service password-encryption
```

```
!
hostname Switch
!
!
!
spanning-tree mode pvst
!
interface FastEthernet0/1
!
interface FastEthernet0/2
!
interface FastEthernet0/3
!
interface FastEthernet0/4
!
interface FastEthernet0/5
!
interface FastEthernet0/6
!
interface FastEthernet0/7
!
interface FastEthernet0/8
!
interface FastEthernet0/9
!
interface FastEthernet0/10
!
interface FastEthernet0/11
!
interface FastEthernet0/12
!
interface FastEthernet0/13
!
interface FastEthernet0/14
!
interface FastEthernet0/15
!
interface FastEthernet0/16
!
interface FastEthernet0/17
!
interface FastEthernet0/18
```

```
!
interface FastEthernet0/19
!
interface FastEthernet0/20
!
interface FastEthernet0/21
!
interface FastEthernet0/22
!
interface FastEthernet0/23
!
interface FastEthernet0/24
!
interface Vlan1
 no ip address
 shutdown
!
!
!
!
line con 0
!
line vty 0 4
 login
line vty 5 15
 login
!
!
end
Switch#configure
Switch(config)#interface fastEthernet 0/15
Switch(config-if)#?
  cdp              Global CDP configuration subcommands
  channel-group    Etherchannel/port bundling configuration
  channel-protocol Select the channel protocol (LACP, PAgP)
  description      Interface specific description
  duplex           Configure duplex operation.
  exit             Exit from interface configuration mode
  ip               Interface Internet Protocol config commands
  mls              mls interface commands
  no               Negate a command or set its defaults
  shutdown         Shutdown the selected interface
```

```
    spanning-tree      Spanning Tree Subsystem
    speed              Configure speed operation.
    storm-control      storm configuration
    switchport         Set switching mode characteristics
    tx-ring-limit      Configure PA level transmit ring limit
Switch(config-if)#ex
Switch(config)#interface vlan 1
Switch(config-if)#?
Interface configuration commands:
    arp            Set arp type (arpa, probe, snap) or timeout
    description    Interface specific description
    exit           Exit from interface configuration mode
    ip             Interface Internet Protocol config commands
    ipv6           IPv6 interface subcommands
    no             Negate a command or set its defaults
    shutdown       Shutdown the selected interface
    standby        HSRP interface configuration commands
Switch(config-if)#ex
Switch(config)#vlan 1
Switch(config-vlan)#exit
Switch(config)#vlan 10
Switch(config-vlan)#exit
Switch(config)#^Z
Switch#
Switch#conf
Switch#configure terminal
Switch(config)#hostname wanglin-S
wanglin-S(config)#exit
wanglin-S#
wanglin-S#
```

实验七　交换机 VLAN 划分实验

一、实验目的

1. 了解 VLAN 原理。
2. 熟练掌握二层交换机 VLAN 的划分方法。
3. 了解如何验证 VLAN 的划分。

二、相关知识

VLAN 是一种逻辑上的网络划分技术,它可以将一个物理网络划分为多个虚拟网络。不同 VLAN 内的主机无法直接通信,必须通过路由器进行相互通信。交换机VLAN划分主要用于提高网络的性能、安全性和可管理性。

交换机 VLAN 划分的应用场景如下。

(1) 校园网络划分:学校实验楼中有两个实验室位于同一楼层,一个是计算机软件实验室,另一个是多媒体实验室,两个实验室的信息端口都连接在一台交换机上。学校已经为实验楼分配了固定的 IP 地址段,为了保证两个实验室的相对独立,需要划分对应的VLAN,使交换机某些端口属于软件实验室,某些端口属于多媒体实验室,这样就能保证它们之间的数据互不干扰,也不影响各自的通信效率。

(2) 企业内部网络划分:根据企业内部的部门、团队或项目划分VLAN,以提高网络安全性,便于企业内部对部门、人员的管理和维护。

(3) 数据中心网络划分:为不同业务系统和服务群划分VLAN,从而提高网络性能并降低网络故障风险。

(4) 多租户网络环境:在数据中心或者园区网络中,为不同租户划分VLAN,确保租户之间的数据隔离。

三、实验设备

1. DCS-3926S 交换机 1 台。
2. PC 机 2 台。
3. Console 线 1 根。
4. 直通网线 2 根。

四、实验拓扑

该实验拓扑结构如图 1-27 所示。

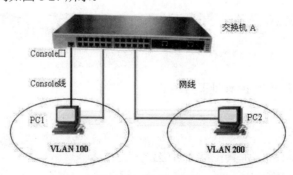

图 1-27 实验拓扑图

使用一台交换机和两台 PC,将其中 PC1 作为控制台终端,使用 Console 口配置方式;使用两根网线分别将 PC1 和 PC2 连接到交换机的 RJ-45 接口上。

五、实验要求

在交换机上划分两个基于端口的 VLAN：VLAN 100 和 VLAN 200。如表 1-2 所示。

表 1-2 VLAN 端口划分

VLAN	端口成员
100	1~8
200	9~16

使得 VLAN 100 的成员能够互相访问，VLAN 200 的成员能够互相访问；VLAN 100 和 VLAN 200 成员之间不能互相访问。

PC1 和 PC2 的网络设置如表 1-3 所示。

表 1-3 设备的 IP 地址分配

设备	IP 地址	子网掩码
交换机 A	192.168.2.11	255.255.255.0
PC1	192.168.2.101	255.255.255.0
PC2	192.168.2.102	255.255.255.0

PC1、PC2 接在 VLAN 100 的成员端口 1~8 上，两台 PC 互相可以 ping 通；PC1、PC2 接在 VLAN 的成员端口 9~16 上，两台 PC 互相可以 ping 通；PC1 接在 VLAN 100 的成员端口 1~8 上，PC2 接在 VLAN 200 的成员端口 9~16 上，则互相 ping 不通。

若实验结果和理论相符，则本实验完成。

六、实验步骤

神州数码设备的配置步骤如下。

第一步：将交换机恢复出厂设置。

```
switch#set default
switch#write
switch#reload
```

第二步：给交换机设置 IP 地址即管理 IP。

```
switch#config
switch(Config)#interface vlan 1
switch(Config-If-Vlan1)#ip address 192.168.2.11 255.255.255.0
switch(Config-If-Vlan1)#no shutdown
switch(Config-If-Vlan1)#exit
switch(Config)#exit
```

第三步：创建 VLAN 100 和 VLAN 200。

switch(Config)#
switch(Config)#vlan 100
switch(Config-Vlan100)#exit
switch(Config)#vlan 200
switch(Config-Vlan200)#exit
switch(Config)#

验证配置：

switch#show vlan				
VLAN Name	Type	Media	Ports	
---- ---------- ------- ------- ----------------------------				
1 default	Static	ENET	Ethernet0/0/1	Ethernet0/0/2
			Ethernet0/0/3	Ethernet0/0/4
			Ethernet0/0/5	Ethernet0/0/6
			Ethernet0/0/7	Ethernet0/0/8
			Ethernet0/0/9	Ethernet0/0/10
			Ethernet0/0/11	Ethernet0/0/12
			Ethernet0/0/13	Ethernet0/0/14
			Ethernet0/0/15	Ethernet0/0/16
			Ethernet0/0/17	Ethernet0/0/18
			Ethernet0/0/19	Ethernet0/0/20
			Ethernet0/0/21	Ethernet0/0/22
			Ethernet0/0/23	Ethernet0/0/24
100 VLAN0100	Static	ENET	！已经创建了 VLAN 100，VLAN 100 中没有端口	
200 VLAN0200	Static	ENET	！已经创建了 VLAN 200，VLAN 200 中没有端口	

第四步：给 VLAN 100 和 VLAN 200 添加端口。

switch(Config)#vlan 100   ！进入 VLAN 100
switch:(Config-Vlan100)#switchport interface ethernet 0/0/1-8   ！给 VLAN 100 加入端口 1~8
Set the port Ethernet0/0/1 access vlan 100 successfully
Set the port Ethernet0/0/2 access vlan 100 successfully
Set the port Ethernet0/0/3 access vlan 100 successfully
Set the port Ethernet0/0/4 access vlan 100 successfully
Set the port Ethernet0/0/5 access vlan 100 successfully
Set the port Ethernet0/0/6 access vlan 100 successfully
Set the port Ethernet0/0/7 access vlan 100 successfully
Set the port Ethernet0/0/8 access vlan 100 successfully
switch(Config-Vlan100)#exit
switch(Config)#vlan 200   ！进入 VLAN 200
switch(Config-Vlan200)#switchport interface ethernet 0/0/9-16   ！为 VLAN 200 加入端口 9~16

Set the port Ethernet0/0/9 access vlan 200 successfully
Set the port Ethernet0/0/10 access vlan 200 successfully
Set the port Ethernet0/0/11 access vlan 200 successfully
Set the port Ethernet0/0/12 access vlan 200 successfully
Set the port Ethernet0/0/13 access vlan 200 successfully
Set the port Ethernet0/0/14 access vlan 200 successfully
Set the port Ethernet0/0/15 access vlan 200 successfully
Set the port Ethernet0/0/16 access vlan 200 successfully
switch(Config-Vlan200)#exit

验证配置：

switch#show vlan

VLAN	Name	Type	Media	Ports	
1	default	Static	ENET	Ethernet0/0/17	Ethernet0/0/18
				Ethernet0/0/19	Ethernet0/0/20
				Ethernet0/0/21	Ethernet0/0/22
				Ethernet0/0/23	Ethernet0/0/24
100	VLAN0100	Static	ENET	Ethernet0/0/1	Ethernet0/0/2
				Ethernet0/0/3	Ethernet0/0/4
				Ethernet0/0/5	Ethernet0/0/6
				Ethernet0/0/7	Ethernet0/0/8
200	VLAN0200	Static	ENET	Ethernet0/0/9	Ethernet0/0/10
				Ethernet0/0/11	Ethernet0/0/12
				Ethernet0/0/13	Ethernet0/0/14
				Ethernet0/0/15	Ethernet0/0/16

思科设备的配置命令如下：

wanglin-S>
wanglin-S>en
wanglin-S#
wanglin-S#configure terminal
Enter configuration commands, one per line. End with CNTL/Z.
wanglin-S(config)#interface vlan 1
wanglin-S(config-if)#ip address 192.168.2.11 255.255.255.0
wanglin-S(config-if)#no shutdown
wanglin-S(config-if)#ex
wanglin-S(config)#vl
wanglin-S(config)#vlan 100
wanglin-S(config-vlan)#ex
wanglin-S(config)#vlan 200
wanglin-S(config-vlan)#ex

```
wanglin-S(config)#interface range fastEthernet 0/1-8
wanglin-S(config-if-range)#switchport access vlan 100
wanglin-S(config-if-range)#exit
wanglin-S(config)#interface range fastEthernet 0/9-16
wanglin-S(config-if-range)#switchport access vlan 200
wanglin-S(config-if-range)#ex
wanglin-S(config)#ex
wanglin-S(config)#exit
wanglin-S#
%SYS-5-CONFIG_I: Configured from console by console
wanglin-S#show vlan
```

VLAN	Name	Status	Ports
1	default	active	Fa0/17, Fa0/18, Fa0/19, Fa0/20
			Fa0/21, Fa0/22, Fa0/23, Fa0/24
100	VLAN0100	active	Fa0/1, Fa0/2, Fa0/3, Fa0/4
			Fa0/5, Fa0/6, Fa0/7, Fa0/8
200	VLAN0200	active	Fa0/9, Fa0/10, Fa0/11, Fa0/12
			Fa0/13, Fa0/14, Fa0/15, Fa0/16
1002	fddi-default	act/unsup	
1003	token-ring-default	act/unsup	
1004	fddinet-default	act/unsup	
1005	trnet-default	act/unsup	

VLAN	Type	SAID	MTU	Parent	RingNo	BridgeNo	Stp	BrdgMode	Trans1	Trans2
1	enet	100001	1500	-	-	-	-	-	0	0
100	enet	100100	1500	-	-	-	-	-	0	0
200	enet	100200	1500	-	-	-	-	-	0	0
1002	fddi	101002	1500	-	-	-	-	-	0	0
1003	tr	101003	1500	-	-	-	-	-	0	0
1004	fdnet	101004	1500	-	-	-	ieee	-	0	0
1005	trnet	101005	1500	-	-	-	ibm	-	0	0

Remote SPAN VLANs

Primary	Secondary	Type	Ports

wanglin-S#

第五步：验证实验。

实验结果如表 1-4 所示。

表1-4 设备 ping 的结果

PC1 位置	PC2 位置	动作	结果
1~8 端口		PC1 ping 192.168.2.11	不通
9~16 端口		PC1 ping 192.168.2.11	不通
17~24 端口		PC1 ping 192.168.2.11	通
1~8 端口	1~8 端口	PC1 ping PC2	通
1~8 端口	9~16 端口	PC1 ping PC2	不通
1~8 端口	17~24 端口	PC1 ping PC2	不通

七、注意事项和排错

1. 默认情况下，交换机所有端口都属于 VLAN 1。我们通常把 VLAN 1 作为交换机的管理 VLAN，因此 VLAN 1 接口的 IP 地址就是交换机的管理地址。

2. 在 DCS-3926S 中，一个普通端口只属于一个 VLAN。

八、配置序列

```
Switch#Show run
Current configuration:
!
    hostname switch
!
Vlan 1
    vlan 1
!
Vlan 100
    vlan 100
!
Vlan 200
    vlan 200
!
Interface Ethernet0/0/1
    switchport access vlan 100
!
Interface Ethernet0/0/2
    switchport access vlan 100
!
Interface Ethernet0/0/3
    switchport access vlan 100
!
```

Interface Ethernet0/0/4
    switchport access vlan 100
!
Interface Ethernet0/0/5
    switchport access vlan 100
!
Interface Ethernet0/0/6
    switchport access vlan 100
!
Interface Ethernet0/0/7
    switchport access vlan 100
!
Interface Ethernet0/0/8
    switchport access vlan 100
!
Interface Ethernet0/0/9
    switchport access vlan 200
!
Interface Ethernet0/0/10
    switchport access vlan 200
!
Interface Ethernet0/0/11
    switchport access vlan 200
!
Interface Ethernet0/0/12
    switchport access vlan 200
!
Interface Ethernet0/0/13
    switchport access vlan 200
!
Interface Ethernet0/0/14
    switchport access vlan 200
!
Interface Ethernet0/0/15
    switchport access vlan 200
!
Interface Ethernet0/0/16
    switchport access vlan 200
!
Interface Ethernet0/0/17
!
Interface Ethernet0/0/18

!
!
Interface Ethernet0/0/24
!
switch#

九、思考题

1. 怎样取消一个 VLAN？
2. 怎样取消一个 VLAN 中的某些端口？

十、课后练习

场景描述：假设你是一家中型企业的网络管理员，公司有以下几个部门：财务部、市场部、技术部。公司现有 1 台交换机(SW1)，为了提高网络安全性、便于管理和优化网络性能，公司决定对网络进行 VLAN 划分。

设计一个 VLAN 划分方案，并简述如何验证 VLAN 划分是否成功。

交换机划分 3 个 VLAN，VLAN 成员表如表 1-5 所示。

表 1-5 VLAN 成员表

VLAN	端口成员
10	1~6
20	7~12
30	13~16

实验八 跨交换机相同 VLAN 间通信

一、实验目的

1. 了解 IEEE802.1q 的实现方法，掌握跨二层交换机相同 VLAN 间通信的调试方法。
2. 了解交换机接口的 Trunk 模式和 Access 模式。
3. 了解交换机的 Tagged 端口和 Untagged 端口的区别。

二、相关知识

跨交换机的相同 VLAN 间通信是网络设计中的常见需求，它允许连接到不同交换机但属于同一 VLAN 的设备进行相互通信。

以下是跨交换机相同 VLAN 间通信的主要应用场景。

(1) 多校区教育网络：在大学或学校网络环境中，不同的教学楼或宿舍楼之间可能需要共享资源。例如，一个教学楼有两层，分别用于一年级和二年级，每个楼层配备一台交换机以满足教师的上网需求。每个年级都有语文教研组和数学教研组，两个年级语文教研组的 PC 可以互相访问，两个年级数学教研组的 PC 也可以互相访问，但语文教研组和数学教研组之间不能自由访问。通过划分 VLAN，可以限制语文教研组和数学教研组之间的访问权限，并使用 IEEE802.1q 协议实现跨交换机的 VLAN 通信。

(2) 大型企业网络：在大型企业中，不同的部门可能分布在不同的建筑或楼层，但需要访问相同的资源。跨交换机的 VLAN 通信能够使不同地点的部门实现无缝连接，共享所需资源。

(3) 数据中心：数据中心通常包含大量分布在不同机架上的服务器，这些服务器可能需要被分配到相同的 VLAN，以实现特定的业务需求。

三、实验设备

1. DCS-3926S 交换机 2 台。
2. PC 机 2 台。
3. Console 线 1 根。
4. 直通网线 2 根。

四、实验拓扑

该实验拓扑结构如图 1-28 所示。

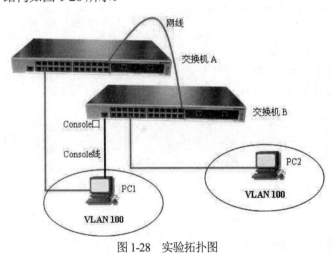

图 1-28　实验拓扑图

五、实验要求

在交换机 A 和交换机 B 上分别划分两个基于端口的 VLAN：VLAN 100 和 VLAN 200，如表 1-6 所示。

表1-6　VLAN成员表

VLAN	端口成员
100	1~8
200	9~16
Trunk	24

使得交换机之间 VLAN 100 的成员能够互相访问，VLAN 200 的成员能够互相访问；VLAN 100 和 VLAN 200 的成员之间不能互相访问。

PC1 和 PC2 的网络设置如表 1-7 所示。

表1-7　设备的 IP 地址分配

设备	IP 地址	子网掩码
交换机 A	192.168.2.11	255.255.255.0
交换机 B	192.168.2.12	255.255.255.0
PC1	192.168.2.101	255.255.255.0
PC2	192.168.2.102	255.255.255.0

PC1 和 PC2 分别接在不同交换机 VLAN 100 的成员端口 1~8 上，两台 PC 互相可以 ping 通；PC1 和 PC2 分别接在不同交换机 VLAN 的成员端口 9~16 上，两台 PC 互相可以 ping 通；PC1 和 PC2 接在不同 VLAN 的成员端口上则互相 ping 不通。

若实验结果和理论相符，则本实验完成。

六、实验步骤

神州数码设备的配置步骤如下。

第一步：将交换机恢复出厂设置。

```
switch#set default
switch#write
switch#reload
```

第二步：给交换机设置标示符和管理 IP。

交换机 A：

```
switch(Config)#hostname switchA
switchA(Config)#interface vlan 1
switchA(Config-If-Vlan1)#ip address 192.168.2.11 255.255.255.0
switchA(Config-If-Vlan1)#no shutdown
switchA(Config-If-Vlan1)#exit
switchA(Config)#
```

交换机 B：

```
switch(Config)#hostname switchB
switchB(Config)#interface vlan 1
switchB(Config-If-Vlan1)#ip address 192.168.2.12 255.255.255.0
switchB(Config-If-Vlan1)#no shutdown
switchB(Config-If-Vlan1)#exit
switchB(Config)#
```

第三步：在交换机中创建 VLAN 100 和 VLAN 200，并添加端口。

交换机 A：

```
switchA(Config)#vlan 100
switchA(Config-Vlan100)#
switchA(Config-Vlan100)#switchport interface ethernet 0/0/1-8
switchA(Config-Vlan100)#exit
switchA(Config)#vlan 200
switchA(Config-Vlan200)#switchport interface ethernet 0/0/9-16
switchA(Config-Vlan200)#exit
switchA(Config)#
```

验证配置：

```
switchA#show vlan
VLAN Name           Type      Media    Ports
1    default        Static    ENET     Ethernet0/0/17    Ethernet0/0/18
                                       Ethernet0/0/19    Ethernet0/0/20
                                       Ethernet0/0/21    Ethernet0/0/22
                                       Ethernet0/0/23    Ethernet0/0/24
100  VLAN0100       Static    ENET     Ethernet0/0/1     Ethernet0/0/2
                                       Ethernet0/0/3     Ethernet0/0/4
                                       Ethernet0/0/5     Ethernet0/0/6
                                       Ethernet0/0/7     Ethernet0/0/8
200  VLAN0200       Static    ENET     Ethernet0/0/9     Ethernet0/0/10
                                       Ethernet0/0/11    Ethernet0/0/12
                                       Ethernet0/0/13    Ethernet0/0/14
                                       Ethernet0/0/15    Ethernet0/0/16
switchA#
```

交换机 B：配置与交换机 A 一样。

第四步：设置交换机 Trunk 端口。

交换机 A：

```
switchA(Config)#interface ethernet 0/0/24
```

switchA(Config-Ethernet0/0/24)#switchport mode trunk
Set the port Ethernet0/0/24 mode TRUNK successfully
switchA(Config-Ethernet0/0/24)#switchport trunk allowed vlan all
set the port Ethernet0/0/24 allowed vlan successfully
switchA(Config-Ethernet0/0/24)#exit
switchA(Config)#

验证配置：

switchA#show vlan

VLAN	Name	Type	Media	Ports	
1	default	Static	ENET	Ethernet0/0/17	Ethernet0/0/18
				Ethernet0/0/19	Ethernet0/0/20
				Ethernet0/0/21	Ethernet0/0/22
				Ethernet0/0/23	Ethernet0/0/24(T)
100	VLAN0100	Static	ENET	Ethernet0/0/1	Ethernet0/0/2
				Ethernet0/0/3	Ethernet0/0/4
				Ethernet0/0/5	Ethernet0/0/6
				Ethernet0/0/7	Ethernet0/0/8
				Ethernet0/0/24(T)	
200	VLAN0200	Static	ENET	Ethernet0/0/9	Ethernet0/0/10
				Ethernet0/0/11	Ethernet0/0/12
				Ethernet0/0/13	Ethernet0/0/14
				Ethernet0/0/15	Ethernet0/0/16
				Ethernet0/0/24(T)	

switchA#

24 口已经出现在 VLAN 1、VLAN 100 和 VLAN 200 中，并且 24 口不是一个普通端口，是 Tagged 端口。

交换机 B：配置与交换机 A 一样。

思科设备的配置命令如下：

S-A 配置

Switch>
Switch>enable
Switch#configure
Configuring from terminal, memory, or network [terminal]?
Enter configuration commands, one per line.　End with CNTL/Z.
Switch(config)#
Switch(config)#hostname S-A
S-A(config)#interface vlan 1

S-A(config-if)#ip address 192.168.2.11 255.255.255.0

S-A(config-if)#no shutdown

S-A(config-if)#exit

S-A(config)#vlan 100

S-A(config-vlan)#exit

S-A(config)#vlan 200

S-A(config-vlan)#exit

S-A(config)#interface range fastEthernet 0/1-8

S-A(config-if-range)#switchport access vlan 100

S-A(config-if-range)#exit

S-A(config)#interface range fastEthernet 0/9-16

S-A(config-if-range)#switchport access vlan 200

S-A(config-if-range)#exit

S-A(config)#interface fastEthernet 0/24

S-A(config-if)#switchport mode trunk

S-A(config-if)#switchport trunk allowed vlan all

S-A(config-if)#exit

S-A(config)#exit

S-A#

%SYS-5-CONFIG_I: Configured from console by console

S-A#show vlan

VLAN	Name	Status	Ports
1	default	active	Fa0/17, Fa0/18, Fa0/19, Fa0/20
			Fa0/21, Fa0/22, Fa0/23
100	VLAN0100	active	Fa0/1, Fa0/2, Fa0/3, Fa0/4
			Fa0/5, Fa0/6, Fa0/7, Fa0/8
200	VLAN0200	active	Fa0/9, Fa0/10, Fa0/11, Fa0/12
			Fa0/13, Fa0/14, Fa0/15, Fa0/16
1002	fddi-default		act/unsup
1003	token-ring-default		act/unsup
1004	fddinet-default		act/unsup
1005	trnet-default		act/unsup

VLAN	Type	SAID	MTU	Parent	RingNo	BridgeNo	Stp	BrdgMode	Trans1	Trans2
1	enet	100001	1500	-	-	-	-	-	0	0
100	enet	100100	1500	-	-	-	-	-	0	0
200	enet	100200	1500	-	-	-	-	-	0	0
1002	fddi	101002	1500	-	-	-	-	-	0	0
1003	tr	101003	1500	-	-	-	-	-	0	0

| 1004 | fdnet | 101004 | 1500 | - | - | - | ieee | - | 0 | 0 |
| 1005 | trnet | 101005 | 1500 | - | - | - | ibm | - | 0 | 0 |

Remote SPAN VLANs

Primary Secondary Type   Ports
------- --------- ---------------- ------------------------------------

S-A#sh
S-A#show run
S-A#show running-config
Building configuration...
Current configuration : 1459 bytes
!
version 12.1
no service timestamps log datetime msec
no service timestamps debug datetime msec
no service password-encryption
!
hostname S-A
!
spanning-tree mode pvst
!
interface FastEthernet0/1
 switchport access vlan 100
!
interface FastEthernet0/2
 switchport access vlan 100
!
interface FastEthernet0/3
 switchport access vlan 100
!
interface FastEthernet0/4
 switchport access vlan 100
!
interface FastEthernet0/5
 switchport access vlan 100
!
interface FastEthernet0/6
 switchport access vlan 100
!
interface FastEthernet0/7
 switchport access vlan 100
!

```
interface FastEthernet0/8
 switchport access vlan 100
!
interface FastEthernet0/9
 switchport access vlan 200
!
interface FastEthernet0/10
 switchport access vlan 200
!
interface FastEthernet0/11
 switchport access vlan 200
!
interface FastEthernet0/12
 switchport access vlan 200
!
interface FastEthernet0/13
 switchport access vlan 200
!
interface FastEthernet0/14
 switchport access vlan 200
!
interface FastEthernet0/15
 switchport access vlan 200
!
interface FastEthernet0/16
 switchport access vlan 200
!
interface FastEthernet0/17
!
interface FastEthernet0/18
!
interface FastEthernet0/19
!
interface FastEthernet0/20
!
interface FastEthernet0/21
!
interface FastEthernet0/22
!
interface FastEthernet0/23
!
interface FastEthernet0/24
```

```
    switchport mode trunk
!
interface Vlan1
  ip address 192.168.2.11 255.255.255.0
!
end
A#
```

交换机 B：配置与交换机 A 一样。

第五步：验证实验。
交换机 A ping 交换机 B：

```
switchA#ping 192.168.2.12
Type ^c to abort.
Sending 5 56-byte ICMP Echos to 192.168.2.12，timeout is 2 seconds.
!!!!!
Success rate is 100 percent (5/5), round-trip min/avg/max = 1/1/1 ms
switchA#
```

表明交换机之前的 Trunk 链路已经成功建立。按表 1-8 验证，PC1 连接在交换机 A 上，PC2 连接在交换机 B 上。

表 1-8 ping 命令的结果

PC1 位置	PC2 位置	动作	结果
1~8 端口		PC1 ping 交换机 B	不通
9~16 端口		PC1 ping 交换机 B	不通
17~24 端口		PC1 ping 交换机 B	通
1~8 端口	1~8 端口	PC1 ping PC2	通
1~8 端口	9~16 端口	PC1 ping PC2	不通

七、注意事项和排错

1. 取消一个 VLAN 可以使用 no vlan 命令。

2. 取消 VLAN 的某个端口可以在 VLAN 模式下使用 no switchport interface ethernet 0/0/x 命令。

3. 当使用 switchport trunk allowed vlan all 命令后，所有之后创建的 VLAN 中都会自动添加 Trunk 端口为成员端口。

八、配置序列

```
switchA#show run
Current configuration:
```

```
!
    hostname switchA
!
Vlan 1
    vlan 1
!
Vlan 100
    vlan 100
!
Vlan 200
    vlan 200
!
!
Interface Ethernet0/0/1
    switchport access vlan 100
!
Interface Ethernet0/0/2
    switchport access vlan 100
!
Interface Ethernet0/0/3
    switchport access vlan 100
!
Interface Ethernet0/0/4
    switchport access vlan 100
!
Interface Ethernet0/0/5
    switchport access vlan 100
!
Interface Ethernet0/0/6
    switchport access vlan 100
!
Interface Ethernet0/0/7
    switchport access vlan 100
!
Interface Ethernet0/0/8
    switchport access vlan 100
!
Interface Ethernet0/0/9
    switchport access vlan 200
!
```

Interface Ethernet0/0/10
    switchport access vlan 200
!
Interface Ethernet0/0/11
    switchport access vlan 200
!
Interface Ethernet0/0/12
    switchport access vlan 200
!
Interface Ethernet0/0/13
    switchport access vlan 200
!
Interface Ethernet0/0/14
    switchport access vlan 200
!
Interface Ethernet0/0/15
    switchport access vlan 200
!
Interface Ethernet0/0/16
    switchport access vlan 200
!
Interface Ethernet0/0/17
!
Interface Ethernet0/0/18
!
Interface Ethernet0/0/19
!
Interface Ethernet0/0/20
!
Interface Ethernet0/0/21
!
Interface Ethernet0/0/22
!
Interface Ethernet0/0/23
!
Interface Ethernet0/0/24
    switchport mode trunk
!
interface Vlan 1
    interface vlan 1
    ip address 192.168.2.11 255.255.255.0

!
switchA#

九、思考题

Trunk、Access、Tagged 和 Untagged 这几个专业术语的关联与区别是什么？

十、课后练习

场景描述：假设你是某大型企业的网络管理员，公司有两个主要办公地点，分别位于不同的建筑物中。每个办公地点都配置有一台核心交换机(分别命名为 SW1 和 SW2)。为了优化网络管理和提高安全性，公司决定实施 VLAN 策略，要求以下部门在两个办公地点之间能够实现跨交换机通信：

- 市场部(VLAN 10)
- 技术部(VLAN 20)
- 财务部(VLAN 30)

所有 VLAN 流量将通过 SW1 和 SW2 之间的 Trunk 链路进行传输。

1. VLAN 划分：列出每个部门的 VLAN 配置。
2. 配置命令：
(1) 写出在 SW1 和 SW2 上创建上述 VLAN 的配置命令。
(2) 写出在 SW1 和 SW2 上将特定端口分配给对应 VLAN 的配置命令。
(3) 写出在 SW1 和 SW2 上配置 Trunk 链路的配置命令。
3. 验证：
(1) 如何验证 VLAN 配置是否正确。
(2) 如何验证 Trunk 链路是否配置正确。

实训项目二 组建简单交换式网络

一、实训目的

1. 能够在多个交换机之间扩展 VLAN，实现更大范围的网络分段。
2. 能够在交换机之间配置 Trunk 链路，通过 Trunk 链路传输多个 VLAN 的数据。
3. 在不同交换机上相同的 VLAN 之间建立通信，确保网络隔离的同时保持必要的连接。
4. 学习规划跨交换机的 VLAN 网络。
5. 能够诊断和解决跨交换机 VLAN 环境中的网络问题。

二、实训设备

(1) DCS-3926S 交换机 2 台。
(2) PC 机 2 台。
(3) 直通网线 3 根。

三、实训拓扑

该实验拓扑结构如图 1-29 所示。

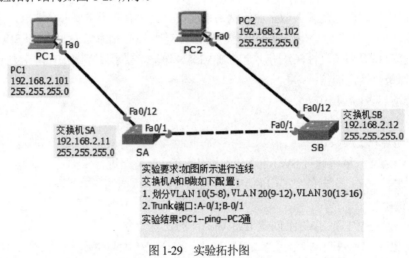

图 1-29　实验拓扑图

四、实训内容

参照图 1-29 所示连线后，PC1 和 PC2 的网络设置如表 1-9 所示。

表 1-9　设备的 IP 地址分配

设备	IP 地址	子网掩码
交换机 A	192.168.2.11	255.255.255.0
交换机 B	192.168.2.12	255.255.255.0
A-PC1	192.168.2.101	255.255.255.0
B-PC2	192.168.2.102	255.255.255.0

PC1 和 PC2 接在 VLAN 10 的成员端口 5~8 上，两台 PC 互相可以 ping 通。
PC1 和 PC2 接在 VLAN 20 的成员端口 9~12 上，两台 PC 互相可以 ping 通。
PC1 和 PC2 接在 VLAN 30 的成员端口 13~16 上，两台 PC 互相可以 ping 通。
PC1 接在 VLAN 10 的成员端口 5~8 上，PC2 接在 VLAN 20 的成员端口 9~12 上，则互相 ping 不通。

若实验结果和理论相符，则本实验完成。
验证实验结果填写在表 1-10 中。

表 1-10 实验结果对照表

PC1 位置	PC2 位置	动作	结果
SA VLAN 10	SB VLAN 10	PC1 ping PC2	
SA VLAN 10	SB VLAN 10	PC1 ping PC2	
SA VLAN 10		PC1 ping SA	
SA VLAN 10		PC1 ping SB	
		SA ping SB	

五、实训步骤

详细实验步骤如下。

(1) 按照实验要求选择对应设备，恢复出厂设置，做常规设置 lachhowww4.2A。

(2) 按照实验要求如图 1-34 所示进行正确连线，注意交换机级联位置 PC1-A12，PC2-B12，A1-B1。

(3) 按要求设置交换机的配置：交换机的地址及管理 IP 设置，VLAN 划分：VLAN 10(5~8)，VLAN 20(9~12)，VLAN 30(13~16)以及级联端口的模式设置 Trunk。

(4) 设置测试主机的 IP 地址。

PC1   192.168.2.101/24
PC2   192.168.2.102/24

(5) 联通测试，使用 ping 命令测试除自己以外的其他网络地址的联通情况，结果与实验要求结果做对比，符合实际情况，实验成功。

(6) PC1 ping PC2 通。

详细配置答案参考以下步骤。

第一步：将交换机恢复出厂设置。

switch#set default
switch#write
switch#reload
switch#la ch

第二步：给交换机设置标示符和管理 IP。

交换机 A：

switch(Config)#hostname ***A
switchA(Config)#interface vlan 1
switchA(Config-If-Vlan1)#ip address 192.168.2.11 255.255.255.0
switchA(Config-If-Vlan1)#no shutdown
switchA(Config-If-Vlan1)#exit
switchA(Config)#

交换机 B：

switch(Config)#hostname ***B
switchB(Config)#interface vlan 1
switchB(Config-If-Vlan1)#ip address 192.168.2.12 255.255.255.0
switchB(Config-If-Vlan1)#no shutdown
switchB(Config-If-Vlan1)#exit
switchB(Config)#

第三步：在交换机中创建 VLAN 100 和 VLAN 200，并添加端口。
神州数码设备的配置命令如下：
交换机 A：

switchA(Config)#vlan 100
switchA(Config-Vlan100)#switchport interface ethernet 0/0/1-8
switchA(Config-Vlan100)#exit
switchA(Config)#vlan 200
switchA(Config-Vlan200)#switchport interface ethernet 0/0/9-16
switchA(Config-Vlan200)#exit
switchA(Config)#

思科设备的配置命令如下：

S-A(config)#vlan 100
S-A(config-vlan)#exit
S-A(config)#vlan 200
S-A(config-vlan)#exit
S-A(config)#interface range fastEthernet 0/1-8
S-A(config-if-range)#switchport access vlan 100
S-A(config-if-range)#exit
S-A(config)#interface range fastEthernet 0/9-16
S-A(config-if-range)#switchport access vlan 200
S-A(config-if-range)#exit

验证配置：

switchA#show vlan

交换机 B：配置与交换机 A 一样。
第四步：设置交换机 Trunk 端口。
交换机 A：

switchA(Config)#interface ethernet 0/0/24
switchA(Config-Ethernet0/0/24)#switchport mode trunk

```
switchA(Config-Ethernet0/0/24)#switchport trunk allowed vlan all
switchA(Config-Ethernet0/0/24)#exit
switchA(Config)#
```

验证配置：

```
switchA#show vlan
```

24 口已经出现在 VLAN 1、VLAN 100 和 VLAN 200 中，并且 24 口不是一个普通端口，是 Tagged 端口。

交换机 B：配置与交换机 A 一样。

第五步：验证实验。

交换机 A ping 交换机 B：

```
switchA#ping 192.168.2.12
Type ^c to abort.
Sending 5 56-byte ICMP Echos to 192.168.2.12, timeout is 2 seconds.
!!!!!
Success rate is 100 percent (5/5), round-trip min/avg/max = 1/1/1 ms
switchA#
```

表明交换机之前的 Trunk 链路已经成功建立。

实验验证结果如表 1-11 所示。

表 1-11 实验结果对照表

PC1 位置	PC2 位置	动作	结果
SA VLAN 10	SB VLAN 10	PC1 ping PC2	通
SA VLAN 10	SB VLAN 10	PC1 ping PC2	不通
SA VLAN 10		PC1 ping SA	不通
SA VLAN 10		PC1 ping SB	不通
		SA ping SB	通

实验九 交换机 MAC 与 IP 的绑定

一、实验目的

1. 理解 MAC 地址与 IP 地址绑定在网络管理中的作用和重要性。
2. 通过 MAC 地址与 IP 地址绑定来实施网络访问控制。
3. 能够在交换机上配置 MAC 地址与 IP 地址绑定。
4. 能够诊断和解决 MAC 地址与 IP 地址绑定配置中的问题。

二、相关知识

交换机 MAC 与 IP 地址绑定是一种网络安全技术，通过将特定的 MAC 地址与 IP 地址关联，确保只有特定的设备可以使用指定的 IP 地址。

交换机 MAC 与 IP 地址绑定的应用场景：学校机房或者网吧等需要固定 IP 地址上网的场所，为了防止用户任意修改 IP 地址而导致 IP 地址冲突，可以使用 MAC 与 IP 地址绑定技术。通过将 MAC、IP 和端口绑定在一起，使用户不能随意修改 IP 地址或更改接入端口，从而使内部网络的管理更加完善。交换机的 AM 功能可以实现 MAC 和 IP 的绑定(AM 全称为 Access Management，访问管理)，AM 利用收到的数据报文信息，例如源 IP 地址和源 MAC 地址，与配置硬件地址池相比较。如果找到匹配，则转发数据包；如果没有匹配，则丢弃数据包。

三、实验设备

1. DCS-3926S 交换机 1 台。
2. PC 机 2 台。
3. Console 线 1~2 根。
4. 直通网线若干。

四、实验拓扑

该实验拓扑结构如图 1-30 所示。

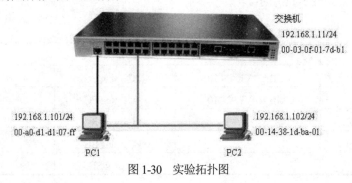

图 1-30　实验拓扑图

五、实验要求

1. 交换机 IP 地址为 192.168.1.11/24，PC1 的地址为 192.168.1.101/24；PC2 的地址为 192.168.1.102/24。
2. 在交换机 0/0/1 端口上将 PC1 的 IP、MAC 与端口绑定。
3. PC1 在 0/0/1 上 ping 交换机的 IP，检验理论是否和实验一致。
4. PC2 在 0/0/1 上 ping 交换机的 IP，检验理论是否和实验一致。
5. PC1 和 PC2 在其他端口上 ping 交换机的 IP，检验理论是否和实验一致。

六、实验步骤

第一步：得到 PC1 主机的 MAC 地址。

```
Microsoft Windows XP [版本 5.1.2600]
(C) 版权所有 1985—2001 Microsoft Corp.
C:\>ipconfig/all
Windows IP Configuration
        Host Name . . . . . . . . . . . : xuxp
        Primary Dns Suffix  . . . . . . : digitalchina.com
        Node Type . . . . . . . . . . . : Broadcast
        IP Routing Enabled. . . . . . . : No
        WINS Proxy Enabled. . . . . . . : No
Ethernet adapter 本地连接:
        Connection-specific DNS Suffix  . :
        Description . . . . . . . . . . : Intel(R) PRO/100 VE Network Connection
        Physical Address. . . . . . . . : 00-A0-D1-D1-07-FF
        Dhcp Enabled. . . . . . . . . . : Yes
        Autoconfiguration Enabled . . . : Yes
        Autoconfiguration IP Address. . . : 169.254.27.232
        Subnet Mask . . . . . . . . . . : 255.255.0.0
        Default Gateway . . . . . . . . :
C:\>
```

得到 PC1 主机的 MAC 地址为：00-A0-D1-D1-07-FF。

第二步：将交换机全部恢复出厂设置，配置交换机的 IP 地址。

```
switch(Config)#interface vlan 1
switch(Config-If-Vlan1)#ip address 192.168.2.11 255.255.255.0
switch(Config-If-Vlan1)#no shut
switch(Config-If-Vlan1)#exit
switch(Config)#
```

第三步：执行 AM 命令。

```
switch(Config)#am enable
switch(Config)#interface ethernet 0/0/1
switch(Config-Ethernet0/0/1)#am mac-ip-pool 00-A0-D1-D1-07-FF 192.168.2.101
switch(Config-Ethernet0/0/1)#exit
```

验证配置：

switch#show am

Am is enabled

Interface Ethernet0/0/1

am mac-ip-pool 00-A0-D1-D1-07-FF 192.168.2.101 USER_CONFIG

第四步：解锁其他端口。

Switch(Config)#interface ethernet 0/0/2

Switch(Config-Ethernet0/0/2)#no am port

Switch(Config)#interface ethernet 0/0/3-20

Switch(Config-Ethernet0/0/3-20)#no am port

第五步：使用 ping 命令验证。

验证结果如表 1-12 所示。

表1-12　ping 命令的结果

PC	端口	Ping	结果	原因
PC1	0/0/1	192.168.2.11	通	
PC1	0/0/7	192.168.2.11	通	
PC2	0/0/1	192.168.2.11	不通	
PC2	0/07	192.168.2.11	通	
PC1	0/0/21	192.168.2.11	不通	
PC2	0/0/21	192.168.2.11	不通	

七、注意事项和排错

1. AM 的默认动作是：拒绝通过(deny)，当 AM 使能的时候，AM 模块会拒绝所有的 IP 报文通过(只允许 IP 地址池内的成员源地址通过)，AM 禁止的时候，AM 会删除所有的地址池。

2. 对于 AM，由于其硬件资源有限，每个 block(8 个端口)最多只能配置 256 条表项。

3. AM 资源要求用户配置的 IP 地址和 MAC 地址不能冲突，也就是说，同一个交换机上不同用户不允许出现相同的 IP 或 MAC 配置。

八、课后练习

场景描述：假设你是某企业网络管理员，负责维护公司内部网络的安全和稳定。近期公司网络出现了一些安全问题，包括 IP 地址冲突和不正当访问。为了提高网络安全性，公司决定实施 MAC 与 IP 地址绑定策略。

1. 网络规划：确定每台设备的地址和 MAC 地址。

2. MAC 和 IP 绑定：

(1) 给需要绑定的设备配置静态 MAC 地址表。

(2) 将 MAC 地址表和其特定的 IP 地址关联。

(3) 配置交换端口，只有绑定的 MAC 地址可以从指定的 IP 地址访问网络。

3. 验证：

(1) 使用已经绑定的设备连接网络，测试并记录结果。

(2) 使用未绑定的设备连接网络，测试并记录结果。

实验十 生成树实验

一、实验目的

1. 掌握生成树协议的基本原理及作用。
2. 在交换机上配置生成树协议。
3. 实验验证生成树协议的配置，确保网络不存在环路。
4. 诊断并解决生成树协议配置中的问题。

二、相关知识

交换机生成树协议(Spanning Tree Protocol, STP)是一种用于防止网络环路的协议。STP 通过在交换机之间建立树状结构，确保只有一个有效的数据传输路径，从而避免环路引发的广播风暴和数据丢失。

IEEE802.1d 协议通过在交换机上运行一套复杂的生成树算法(Spanning Tree Algorithm，STA)，将冗余端口置于"阻断状态"，确保 PC 在与其他 PC 通信时，只有一条链路生效，如果该链路出现故障无法使用，IEEE802.1d 协议会重新计算网络链路，将处于"阻断状态"的端口重新激活，从而既保障了网络的正常运行，又保留了冗余能力。

生成树协议主要应用于以下场景：

(1) 防止网络环路：在企业或教育网络中，通过配置生成树协议可以有效防止网络环路，从而确保网络的稳定性和可靠性。

(2) 网络扩展和升级：在进行网络扩展或升级时，生成树协议可以帮助网络管理员更好地规划和管理网络拓扑。

三、实验设备

1. DCS-3926S 交换机 2 台。
2. PC 机 2 台。
3. Console 线 1~2 根。
4. 直通网线 4~8 根。

四、实验拓扑

该实验拓扑结构如图 1-31 所示。

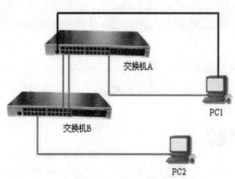

图 1-31 实验拓扑图

五、实验要求

实验中各设备的 IP 地址设置如表 1-13 所示。

表 1-13 IP 地址设置

设备	IP	Mask
交换机 A	192.168.2.11	255.255.255.0
交换机 B	192.168.2.12	255.255.255.0
PC1	192.168.2.101	255.255.255.0
PC2	192.168.2.102	255.255.255.0

网线连接如表 1-14 所示。

表 1-14 网线连接

交换机 A e0/0/1	交换机 B e0/0/3
交换机 A e0/0/2	交换机 B e0/0/4
PC1	交换机 A e0/0/24
PC2	交换机 B e0/0/23

如果生成树成功,则 PC1 可以 ping 通 PC2。

六、实验步骤

神州数码设备和思科设备的配置命令相同,具体步骤如下。

第一步:正确连接网线,恢复出厂设置之后,做初始配置。

交换机 A:

```
switch#config
```

```
switch(Config)#hostname switchA
switchA(Config)#interface vlan 1
switchA(Config-If-Vlan1)#ip address 192.168.2.11 255.255.255.0
switchA(Config-If-Vlan1)#no shutdown
switchA(Config-If-Vlan1)#exit
switchA(Config)#
```

交换机 B：

```
switch#config
switch(Config)#hostname switchB
switchB(Config)#interface vlan 1
switchB(Config-If-Vlan1)#ip address 192.168.2.12   255.255.255.0
switchB(Config-If-Vlan1)#no shutdown
switchB(Config-If-Vlan1)#exit
switchB(Config)#
```

第二步："PC1 ping PC2 -t"观察现象。

1. ping 不通。

2. 所有连接网线的端口的绿灯很频繁地闪烁，表明该端口收发数据量很大，已经在交换机内部形成广播风暴。

第三步：在两台交换机中都启用生成树协议。

```
switchA(Config)#spanning-tree
MSTP is starting now, please wait..........
MSTP is enabled successfully.
switchA(Config)#
switchB(Config)#spanning-tree
MSTP is starting now, please wait..........
MSTP is enabled successfully.
switchB(Config)#
```

验证配置：

```
switchA#show spanning-tree
                -- MSTP Bridge Config Info --
Standard      :   IEEE 802.1s
Bridge MAC    :   00:03:0f:01:25:28
Bridge Times  :   Max Age 20, Hello Time 2, Forward Delay 15
Force Version:   3
```

######################### Instance 0 #########################

Self Bridge Id   : 32768 -   00:03:0f:01:25:28

Root Id   : this switch

Ext.RootPathCost : 0

Region Root Id   : this switch

Int.RootPathCost : 0

Root Port ID   : 0

Current port list in Instance 0:

Ethernet0/0/1 Ethernet0/0/2 (Total 2)

PortName	ID	ExtRPC	IntRPC	State	Role	DsgBridge	DsgPort
Ethernet0/0/1	128.001	0	0	FWD	DSGN	32768.00030f012528	128.001
Ethernet0/0/2	128.002	0	0	FWD	DSGN	32768.00030f012528	128.002

switchA#

switchB#show spanning-tree

-- MSTP Bridge Config Info --

Standard   : IEEE 802.1s

Bridge MAC   : 00:03:0f:01:7d:b0

Bridge Times :   Max Age 20, Hello Time 2, Forward Delay 15

Force Version: 3

######################### Instance 0 #########################

Self Bridge Id   : 32768 -   00:03:0f:01:7d:b0

Root Id   : 32768.00:03:0f:01:25:28

Ext.RootPathCost : 200000

Region Root Id   : this switch

Int.RootPathCost : 0

Root Port ID   : 128.4

Current port list in Instance 0:

Ethernet0/0/3 Ethernet0/0/4 Ethernet0/0/23 (Total 3)

PortName	ID	ExtRPC	IntRPC	State	Role	DsgBridge	DsgPort
Ethernet0/0/3	128.003	0	0	BLK	ALTR	32768.00030f012528	128.002
Ethernet0/0/4	128.004	0	0	FWD	ROOT	32768.00030f012528	128.001
Ethernet0/0/23	128.023	200000	0	FWD	DSGN	32768.00030f017db0	128.023

从 show 命令的输出中可以看出，交换机 A 是根交换机，而交换机 B 的 4 端口是根端口。

第四步：继续使用"PC1 ping PC2 -t"观察现象。
1. 拔掉交换机 B 端口 4 的网线，观察现象，写在下方。
2. 再插上交换机 B 端口 4 的网线，观察现象，写在下方。

七、注意事项和排错

1. 如果想在交换机上运行 MSTP，首先必须在全局打开 MSTP 开关。在没有打开全局 MSTP 开关之前，打开端口的 MSTP 开关是不允许的。

2. MSTP 定时器参数之间是有相关性的，错误配置可能导致交换机不能正常工作。各定时器之间的关联关系为：

(1) 2×(Bridge_Forward_Delay – 1.0 seconds) >= Bridge_Max_Age。

(2) Bridge_Max_Age >= 2×(Bridge_Hello_Time + 1.0 seconds)。

3. 用户在修改 MSTP 参数时，应该清楚所产生的各个拓扑。除全局的基于网桥的参数配置外，其他的是基于各个实例的配置，在配置时一定要注意配置参数对应的实例是否正确。

4. DCS-3926S 交换机的端口 MSTP 功能与端口 MAC 绑定、802.1x 和设置端口为路由端口功能互斥。当端口已经配置 MAC 绑定、802.1x 或设置为路由端口时，无法在该端口启动 MSTP 功能。

八、思考题

1. 生成树协议怎样选取根端口和指定端口？
2. MSTP 通过怎样的策略可以使备份链路实现快速启用？

九、课后练习

场景描述：作为网络管理员，你负责维护公司两个主要办公地点之间的网络连接。为了确保网络的可靠性和冗余性，你决定使用两台交换机(SW1 和 SW2)通过 4 根网线连接，形成一个环形拓扑。你需要配置多生成树协议(MSTP)来管理网络中的冗余链路，并观察备份链路的启用和断开过程。

1. 网络连接：使用 4 根网线将 SW1 和 SW2 连接，形成环形拓扑。
2. 配置 MSTP：在 SW1 和 SW2 上启用 MSTP。
3. 观察根端口：使用 show spanning-tree 命令观察并记录端口的选择情况。
4. 启动备份链路：在 SW1 和 SW2 上启用 debug 功能，观察备份链路的信息。
5. 备份链路测试：使用计时器记录备份链路启用和恢复的时间。

实验十一　交换机链路聚合

一、实验目的

1. 掌握链路聚合的基本原理和作用。
2. 在交换机上配置链路聚合。
3. 实验验证链路聚合的配置。
4. 诊断并解决链路聚合配置中的问题。

二、相关知识

交换机链路聚合是一种技术，它将多个物理链路捆绑在一起，形成一个逻辑链路，以提高带宽、冗余和网络可靠性。链路聚合支持多种模式，包括负载均衡和故障转移。负载均衡模式下，流量会在链路聚合组(LAG)的成员链路之间均匀分配；故障转移模式下，当一条链路出现故障时，其他链路将接管其工作。

交换机链路聚合应用场景如下。

(1) 提高带宽：假如两个实验室分别使用一台交换机提供 20 多个信息点，两个实验室的互通通过一根级联网线。每个实验室的信息点都是百兆到桌面。两个实验室之间的带宽也是 100Mb/s，如果实验室之间需要大量传输数据，就会明显感觉带宽资源紧张。当楼层之间大量用户都希望以 100Mb/s 传输数据时，楼层间的链路就呈现出了独木桥的状态，必然造成网络传输效率下降等后果。解决这个问题的办法就是提高楼层主交换机之间的连接带宽，实现的办法可以是采用千兆端口替换原来的 100Mb/s 端口进行互联，但这样无疑会增加组网的成本，需要更新端口模块，并且线缆也需要进行进一步的升级。另一种相对经济的升级方案是采用链路聚合技术。通过将 4 条 100Mb/s 链路聚合为一个逻辑链路，在全双工模式下可以实现高达 800Mb/s 的带宽，接近千兆的传输能力。这种方式不仅成本较低，实施起来也相对简便。

(2) 提供冗余：在需要高可靠性的应用场景中(如服务器连接或关键网络链路)，链路聚合可以提供冗余，确保网络不中断。

(3) 网络扩展与升级：在进行网络扩展或升级时，链路聚合技术可以帮助网络管理员更好地规划和管理网络拓扑。

三、实验设备

1. DCS-3926S 交换机 2 台。
2. PC 机 2 台。
3. Console 线 1~2 根。
4. 直通网线 4~8 根。

四、实验拓扑

交换机链路聚合的实验拓扑图如图 1-32 所示。

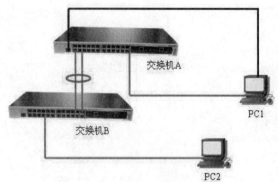

图 1-32　交换机链路聚合实验拓扑图

五、实验要求

设备 IP 配置如表 1-15 所示。

表 1-15　实验配置

设备	IP	Mask	端口
交换机 A	192.168.2.11	255.255.255.0	0/0/1-2 trunking
交换机 B	192.168.2.12	255.255.255.0	0/0/1-2 trunking
PC1	192.168.2.101	255.255.255.0	交换机 A0/0/23
PC2	192.168.2.102	255.255.255.0	交换机 B0/0/24

如果链路聚合成功，则 PC1 可以 ping 通 PC2。

六、实验步骤

神州数码设备的配置步骤如下。

第一步：正确连接网线，交换机全部恢复出厂设置，进行初始配置，避免广播风暴出现。

交换机 A：

```
switch#config
switch(Config)#hostname switchA
switchA(Config)#interface vlan 1
switchA(Config-If-Vlan1)#ip address 192.168.2.11 255.255.255.0
switchA(Config-If-Vlan1)#no shutdown
switchA(Config-If-Vlan1)#exit
switchA(Config)#spanning-tree
MSTP is starting now, please wait..........
MSTP is enabled successfully.
```

switchA(Config)#

交换机 B：

switch#config
switch(Config)#hostname switchB
switchB(Config)#interface vlan 1
switchB(Config-If-Vlan1)#ip address 192.168.2.12 255.255.255.0
switchB(Config-If-Vlan1)#no shutdown
switchB(Config-If-Vlan1)#exit
switchB(Config)#spanning-tree
MSTP is starting now, please wait..........
MSTP is enabled successfully.
switchB(Config)#

第二步：创建 port group。
交换机 A：

switchA(Config)#port-group 1
switchA(Config)#

验证配置：

switchA#show port-group detail
Sorted by the ports in the group 1:
----
switchA#show port-group brief
Port-group number : 1
Number of ports in port-group : 0   Maxports in port-channel = 8
Number of port-channels : 0   Max port-channels : 1
switchA#

交换机 B：

switchB(Config)#port-group 2
switchB(Config)#

第三步：手动生成链路聚合组(第三、第四步任选其一操作)。
交换机 A：

switchA(Config)#interface ethernet 0/0/1-2
switchA(Config-Port-Range)#port-group 1 mode on
switchA(Config-Port-Range)#exit
switchA(Config)#interface port-channel 1
switchA(Config-If-Port-Channel1)#

验证配置：

switchA#show vlan

VLAN	Name	Type	Media	Ports	
1	default	Static	ENET	Ethernet0/0/3	Ethernet0/0/4
				Ethernet0/0/5	Ethernet0/0/6
				Ethernet0/0/7	Ethernet0/0/8
				Ethernet0/0/9	Ethernet0/0/10
				Ethernet0/0/11	Ethernet0/0/12
				Ethernet0/0/13	Ethernet0/0/14
				Ethernet0/0/15	Ethernet0/0/16
				Ethernet0/0/17	Ethernet0/0/18
				Ethernet0/0/19	Ethernet0/0/20
				Ethernet0/0/21	Ethernet0/0/22
				Ethernet0/0/23	Ethernet0/0/24
				Port-Channel1	

switchA#   ！port-channel1 已经存在

交换机 B：

```
switchB(Config)#int e 0/0/3-4
switchB(Config-Port-Range)#port-group 2 mode on
switchB(Config-Port-Range)#exit
switchB(Config)#interface port-channel 2
switchB(Config-If-Port-Channel2)#
```

验证配置：

```
switchB#show port-group brief
Port-group number : 2
Number of ports in port-group : 2    Maxports in port-channel = 8
Number of port-channels : 1    Max port-channels : 1
switchB#
```

第四步：LACP 动态生成链路聚合组(第三、第四步任选其一操作)。

```
switchA(Config)#interface ethernet 0/0/1-2
switchA(Conifg-Port-Range)#port-group 1 mode active
switchA(Config)#interface port-channel 1
switchA(Config-If-Port-Channel1)#
```

验证配置：

switchA#show vlan

VLAN	Name	Type	Media	Ports	
1	default	Static	ENET	Ethernet0/0/3	Ethernet0/0/4
				Ethernet0/0/5	Ethernet0/0/6
				Ethernet0/0/7	Ethernet0/0/8
				Ethernet0/0/9	Ethernet0/0/10

	Ethernet0/0/11	Ethernet0/0/12
	Ethernet0/0/13	Ethernet0/0/14
	Ethernet0/0/15	Ethernet0/0/16
	Ethernet0/0/17	Ethernet0/0/18
	Ethernet0/0/19	Ethernet0/0/20
	Ethernet0/0/21	Ethernet0/0/22
	Ethernet0/0/23	Ethernet0/0/24
	Port-Channel1	
switchA#	！port-channel1 已经存在	

交换机 B：

switchB(Config)#interface ethernet 0/0/3-4
switchB(Conifg-Port-Range)#port-group 2 mode passive
switchB(Config)#interface port-channel 2
switchB(Config-If-Port-Channel2)#

验证配置：

switchB#show port-group brief
Port-group number : 2
Number of ports in port-group : 2   Maxports in port-channel = 8
Number of port-channels : 1   Max port-channels : 1
switchB#

思科设备的配置命令如下：

Switch>enable
Switch#configure terminal
Switch(config)#hostname S-A
S-A(config)#
%LINK-5-CHANGED: Interface FastEthernet0/1, changed state to up
%LINEPROTO-5-UPDOWN: Line protocol on Interface FastEthernet0/1, changed state to up
%LINK-5-CHANGED: Interface FastEthernet0/2, changed state to up
%LINEPROTO-5-UPDOWN: Line protocol on Interface FastEthernet0/2, changed state to up
S-A(config)#
S-A(config)#sp
S-A(config)#spanning-tree
% Incomplete command.
S-A(config)#interface vlan 1
S-A(config-if)#ip address 192.168.2.11 255.255.255.0
S-A(config-if)#no shutdown
S-A(config-if)#exit
Switch>enable
Switch#configure terminal
Switch(config)#hostname S-B
S-B(config)#

```
%LINK-5-CHANGED: Interface FastEthernet0/3, changed state to up
%LINEPROTO-5-UPDOWN: Line protocol on Interface FastEthernet0/3, changed state to up
%LINK-5-CHANGED: Interface FastEthernet0/4, changed state to up
%LINEPROTO-5-UPDOWN: Line protocol on Interface FastEthernet0/4, changed state to up
S-B(config)#
S-B(config)#spanning-tree
% Incomplete command.
S-B(config)#
S-B(config)#interface vlan 1
S-B(config-if)#ip address 192.168.2.12 255.255.255.0
S-B(config-if)#no shutdown
S-B(config-if)#exit
```

方案一：通过 LACP 协议动态生成链路聚合组(使用 active 和 passive 聚合参数)，可以实现较高的配合成功率。

```
S-A(config)#interface fastEthernet 0/1
S-A(config-if)#channel-group 1 mode active
S-A(config-if)#interface fastEthernet 0/2
S-A(config-if)#channel-group 1 mode active
S-A(config-if)#interface port-channel 1
S-A#
S-A#sh vlan
VLAN Name                             Status    Ports
---- -------------------------------- --------- -------------------------------
1    default                          active    Fa0/1, Fa0/2, Fa0/3, Fa0/4
                                                Fa0/5, Fa0/6, Fa0/7, Fa0/8
                                                Fa0/9, Fa0/10, Fa0/11, Fa0/12
                                                Fa0/13, Fa0/14, Fa0/15, Fa0/16
                                                Fa0/17, Fa0/18, Fa0/19, Fa0/20
                                                Fa0/21, Fa0/22, Fa0/23, Fa0/24
                                                Po1
1002 fddi-default                     act/unsup
1003 token-ring-default               act/unsup
1004 fddinet-default                  act/unsup
1005 trnet-default                    act/unsup
S-A#show  etherchannel
                Channel-group listing:
                ----------------------

Group: 1
---------
Group state = L2
Ports: 2   Maxports = 16
Port-channels: 1 Max Port-channels = 16
Protocol:    LACP
S-A#sh run
```

```
Building configuration...
Current configuration : 1073 bytes
!
version 12.1
no service timestamps log datetime msec
no service timestamps debug datetime msec
no service password-encryption
!
hostname S-A
!
!
!
spanning-tree mode pvst
!
interface FastEthernet0/1
  channel-group 1 mode active
!
interface FastEthernet0/2
  channel-group 1 mode active
!
interface FastEthernet0/3
!
........
interface FastEthernet0/24
!
interface Port-channel 1
!
interface Vlan1
  ip address 192.168.2.11 255.255.255.0
!
!
!
!
line con 0
!
line vty 0 4
  login
line vty 5 15
  login
!
!
end
S-B(config)#interface range fastEthernet 0/3-4
S-B(config-if-range)#channel-group 2 mode passive
S-B(config-if-range)#
Creating a port-channel interface Port-channel 2
```

```
%LINEPROTO-5-UPDOWN: Line protocol on Interface FastEthernet0/3, changed state to down
%LINEPROTO-5-UPDOWN: Line protocol on Interface FastEthernet0/3, changed state to up
%LINEPROTO-5-UPDOWN: Line protocol on Interface FastEthernet0/4, changed state to down
%LINEPROTO-5-UPDOWN: Line protocol on Interface FastEthernet0/4, changed state to up
S-B(config-if-range)#
S-B(config-if-range)#^Z
S-B#
S-B#sh vlan

VLAN Name                             Status    Ports
---- -------------------------------- --------- -------------------------------
1    default                          active    Fa0/1, Fa0/2, Fa0/3, Fa0/4
                                                Fa0/5, Fa0/6, Fa0/7, Fa0/8
                                                Fa0/9, Fa0/10, Fa0/11, Fa0/12
                                                Fa0/13, Fa0/14, Fa0/15, Fa0/16
                                                Fa0/17, Fa0/18, Fa0/19, Fa0/20
                                                Fa0/21, Fa0/22, Fa0/23, Fa0/24
                                                Po2
1002 fddi-default                     act/unsup
1003 token-ring-default               act/unsup
1004 fddinet-default                  act/unsup
1005 trnet-default                    act/unsup
S-B(config-if)#interface port-channel 2
S-B#show etherchannel
                Channel-group listing:
                ----------------------

Group: 2
---------
Group state = L2
Ports: 2   Maxports = 16
Port-channels: 1 Max Port-channels = 16
Protocol:    LACP
S-B#
S-B#sh run
Building configuration...
Current configuration : 1075 bytes
!
version 12.1
no service timestamps log datetime msec
no service timestamps debug datetime msec
no service password-encryption
!
hostname S-B
!
!
!
```

```
spanning-tree mode pvst
!
interface FastEthernet0/1
!
interface FastEthernet0/2
!
interface FastEthernet0/3
  channel-group 2 mode passive
!
interface FastEthernet0/4
  channel-group 2 mode passive
!
interface FastEthernet0/5
!
interface FastEthernet0/6
!
……
!
interface FastEthernet0/24
!
interface Port-channel 2
!
interface Vlan1
  ip address 192.168.2.12 255.255.255.0
!
!
!
line con 0
!
line vty 0 4
  login
line vty 5 15
  login
!
!
end
```

方案二：通过手动强制方式生成链路聚合组(使用 on 聚合参数)，可以实现较高的配合成功率。

删除已经聚合的端口：

```
S-A# conf
S-A# configure
S-A(config)#no interface Port-channel 1
S-A(config)#^Z
```

```
S-A#
S-B#conf
S-B#configure
S-B(config)#no interface Port-channel 2
S-A#conf
S-A#configure
S-A(config)#interface range fastEthernet 0/1-2
S-A(config-if-range)#channel-group 1 mode on
S-A(config-if-range)#
Creating a port-channel interface Port-channel 1

%LINK-5-CHANGED: Interface Port-channel 1, changed state to up
%LINEPROTO-5-UPDOWN: Line protocol on Interface Port-channel 1, changed state to up
%LINEPROTO-5-UPDOWN: Line protocol on Interface FastEthernet0/1, changed state to down
%LINEPROTO-5-UPDOWN: Line protocol on Interface FastEthernet0/1, changed state to up
%LINEPROTO-5-UPDOWN: Line protocol on Interface FastEthernet0/2, changed state to down
%LINEPROTO-5-UPDOWN: Line protocol on Interface FastEthernet0/2, changed state to up
S-A(config-if-range)#
S-A(config-if-range)#^Z
S-A#
%SYS-5-CONFIG_I: Configured from console by console
S-A#
S-A#sh vl
S-A#sh vlan
```

VLAN	Name	Status	Ports
1	default	active	Fa0/1, Fa0/2, Fa0/3, Fa0/4
			Fa0/5, Fa0/6, Fa0/7, Fa0/8
			Fa0/9, Fa0/10, Fa0/11, Fa0/12
			Fa0/13, Fa0/14, Fa0/15, Fa0/16
			Fa0/17, Fa0/18, Fa0/19, Fa0/20
			Fa0/21, Fa0/22, Fa0/23, Fa0/24
			Po1
1002	fddi-default	act/unsup	
1003	token-ring-default	act/unsup	
1004	fddinet-default	act/unsup	
1005	trnet-default	act/unsup	

VLAN	Type	SAID	MTU	Parent	RingNo	BridgeNo	Stp	BrdgMode	Trans1	Trans2
1	enet	100001	1500	-	-	-	-	-	0	0
1002	fddi	101002	1500	-	-	-	-	-	0	0
1003	tr	101003	1500	-	-	-	-	-	0	0
1004	fdnet	101004	1500	-	-	-	ieee	-	0	0
1005	trnet	101005	1500	-	-	-	ibm	-	0	0

Remote SPAN VLANs

```
---------------------------------------------------------------------
Primary   Secondary    Type              Ports
-------   ---------    --------------    ---------------------------
S-A#sh run
Building configuration...
Current configuration : 1065 bytes
!
version 12.1
no service timestamps log datetime msec
no service timestamps debug datetime msec
no service password-encryption
!
hostname S-A
!
!
!
spanning-tree mode pvst
!
interface FastEthernet0/1
  channel-group 1 mode on
!
interface FastEthernet0/2
  channel-group 1 mode on
!
interface FastEthernet0/3
!
......
!
interface FastEthernet0/24
!
interface Port-channel 1
!
interface Vlan1
  ip address 192.168.2.11 255.255.255.0
!
!
end
S-A#
S-A#
S-B#conf
S-B#configure
S-B(config)#
S-B(config)#interface ran fastEthernet 0/3-4
S-B(config-if-range)#channel-group 2 mode on
S-B(config-if-range)#
```

Creating a port-channel interface Port-channel 2
%LINK-5-CHANGED: Interface Port-channel 2, changed state to up
%LINEPROTO-5-UPDOWN: Line protocol on Interface Port-channel 2, changed state to up
%LINEPROTO-5-UPDOWN: Line protocol on Interface FastEthernet0/3, changed state to down
%LINEPROTO-5-UPDOWN: Line protocol on Interface FastEthernet0/3, changed state to up
%LINEPROTO-5-UPDOWN: Line protocol on Interface FastEthernet0/4, changed state to down
%LINEPROTO-5-UPDOWN: Line protocol on Interface FastEthernet0/4, changed state to up
S-B(config-if-range)#^Z
S-B#
%SYS-5-CONFIG_I: Configured from console by console
S-B#
S-B#sh vl
S-B#sh vlan

VLAN	Name	Status	Ports
1	default	active	Fa0/1, Fa0/2, Fa0/3, Fa0/4
			Fa0/5, Fa0/6, Fa0/7, Fa0/8
			Fa0/9, Fa0/10, Fa0/11, Fa0/12
			Fa0/13, Fa0/14, Fa0/15, Fa0/16
			Fa0/17, Fa0/18, Fa0/19, Fa0/20
			Fa0/21, Fa0/22, Fa0/23, Fa0/24
			Po2
1002	fddi-default	act/unsup	
1003	token-ring-default	act/unsup	
1004	fddinet-default	act/unsup	
1005	trnet-default	act/unsup	

VLAN	Type	SAID	MTU	Parent	RingNo	BridgeNo	Stp	BrdgMode	Trans1	Trans2
1	enet	100001	1500	-	-	-	-	-	0	0
1002	fddi	101002	1500	-	-	-	-	-	0	0
1003	tr	101003	1500	-	-	-	-	-	0	0
1004	fdnet	101004	1500	-	-	-	ieee	-	0	0
1005	trnet	101005	1500	-	-	-	ibm	-	0	0

Remote SPAN VLANs

Primary Secondary Type   Ports
------- --------- ---------------- ------------------------------------

S-B#sh run
Building configuration...
Current configuration : 1065 bytes
!
version 12.1
no service timestamps log datetime msec
no service timestamps debug datetime msec

```
no service password-encryption
!
hostname S-B
!
!
!
spanning-tree mode pvst
!
interface FastEthernet0/1
!
interface FastEthernet0/2
!
interface FastEthernet0/3
  channel-group 2 mode on
!
interface FastEthernet0/4
  channel-group 2 mode on
!
interface FastEthernet0/5
!
.
interface FastEthernet0/24
!
interface Port-channel 2
!
interface Vlan1
  ip address 192.168.2.12 255.255.255.0
!
!
end
S-B#
```

第五步：参照表1-16所示，使用 ping 命令进行验证(使用 PC1 ping PC2)。

表1-16 测试配置表

交换机 A	交换机 B	结果	原因
0/0/1 0/0/2	0/0/3 0/0/4	通	链路聚合组连接正确
0/0/1 0/0/2	0/0/3	通	拔掉交换机 B 端口 4 的网线，仍然可以通(需要一点时间)，此时使用 show vlan 命令查看结果，port-channel 消失。只有一个端口连接时，没有必要再维持一个 port-channel
0/0/1 0/0/2	0/0/5 0/0/6	通	等候一小段时间后，仍然是通的。使用 show vlan 命令查看结果。如果将两台交换机的 spanning-tree 功能 disable，这时候执行第三步和第四步的结果会不同。执行第四步，将会形成环路

七、注意事项和排错

1. 为使 Port Channel 正常工作，Port Channel 的成员端口必须具备以下相同的属性。
(1) 端口均为全双工模式。
(2) 端口速率相同。
(3) 端口的类型必须一样，比如同为以太口或同为光纤口。
(4) 端口同为 Access 端口并且属于同一个 VLAN 或同为 Trunk 端口。
(5) 如果端口为 Trunk 端口，则其 Allowed VLAN 和 Native VLAN 属性也应该相同。
2. 支持任意两个交换机物理端口的汇聚，最大组数为 6 个，组内最多的端口数为 8 个。
3. 一些命令不能在 port-channel 上的端口使用，包括 arp、bandwidth、ip、ip-forward 等。
4. 在使用强制生成端口聚合组时，由于汇聚是手动配置触发的，如果由于端口的 VLAN 信息不一致导致汇聚失败的话，汇聚组一直会停留在没有汇聚的状态，必须通过往该 group 增加和删除端口来触发端口再次汇聚，如果 VLAN 信息还是不一致仍然不能汇聚成功，直到 VLAN 信息都一致并且有增加和删除端口触发汇聚的情况下端口才能汇聚成功。
5. 检查对端交换机的对应端口是否配置端口聚合组，且要查看配置方式是否相同，如果本端是手动方式则对端也应该配置成手动方式，如果本端是 LACP 动态生成则对端也应该是 LACP 动态生成，否则端口聚合组不能正常工作；还有一点要注意的是，如果两端收发的都是 LACP 协议，至少有一端是 Active 的，否则两端都不会发起 LACP 数据报。
6. port-channel 一旦形成之后，所有对应端口的设置只能在 port-channel 端口上进行。
7. LACP 必须和 Security 和 802.1X 的端口互斥，如果端口已经配置上述两种协议，就不允许被启用 LACP。

八、课后练习

场景描述：假设你是一家小型企业的网络管理员，公司有三个办公地点，分别位于不同的建筑中。为了优化网络管理和提高带宽，公司决定实施链路聚合。你需要确保以下部门在各自办公地点之间能够通过交换机实现通信，并实施一些高级网络特性以提升网络性能和可靠性：

- 市场部(VLAN 10)
- 技术部(VLAN 20)
- 财务部(VLAN 30)

1. 链路聚合规划与设计：设计一个网络拓扑图，展示三个办公地点的交换机连接方式，包括交换机之间的 Trunk 链路，并选择合适的链路聚合模式。
2. 高级网络特性配置：在所有交换机上实施 STP(生成树协议)，以防止网络环路。
3. 交换机配置：
(1) 在三个地点的交换机上创建相应的 VLAN。
(2) 将特定端口分配到对应的 VLAN。
(3) 配置交换机之间的 Trunk 链路。

(4) 配置链路聚合。
(5) 配置 STP。
4. 实验验证:
(1) 验证 VLAN 配置是否正确。
(2) 验证链路聚合和 STP 配置是否正确。

实训项目三 基于 VLAN 的交换机链路聚合

一、实训目的

1. 使用两根网线做链路聚合,通过插拔线缆观察结果。
2. 把组作为交换机之间的 Trunk 链路,实现跨交换机的 VLAN。

二、实训设备

(1) DCS-3926S 交换机 2 台。
(2) PC 机 2 台。
(3) 直通网线 4 根。

三、实训拓扑

该实验拓扑结构如图 1-33 所示。

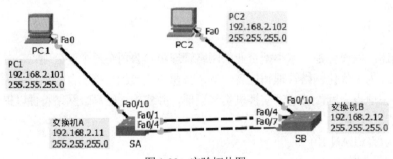

图 1-33 实验拓扑图

四、实训要求

1. 参照如图 1-33 所示进行连线。交换机 A 和 B 做如下配置。
(1) 在交换机 A 和交换机 B 上启用生成树协议。
(2) 划分 VLAN 10(9~15)、VLAN 20(16~24)。
(3) 划分 Trunk 端口 A-0/0/1、A-0/0/4、B-0/0/4、B-0/0/7。
(4) 做链路聚合 A-14 和 B-47,分别聚合生成聚合链路,此链路将作为两台交换机的 Trunk 链路。

2. 实验结果:PC1---ping---PC2　　通。

五、实训步骤

详细步骤说明:

(1) 选择相应设备,并参考图 1-38 判断是否需要优先配置生成树协议;在设备配置模式下,输入命令:SA(config)#sp。

(2) 按要求连线。

4 条: 　 A1---B4　 A4---B7　 PC1---A 0/0/10　 PC2---B 0/0/10

(3) 配置相对应 IP 地址。

设备	IP 地址	Mask
交换机 A	192.168.2.11	255.255.255.0
交换机 B	192.168.2.12	255.255.255.0
A-PC1	192.168.2.101	255.255.255.0
B-PC2	192.168.2.102	255.255.255.0

(4) 测试连通状态。

A—PC1　 ping 192.168.2.11 -t　 ping 192.168.2.12 -t　 ping 192.168.2.102 -t
B—PC2　 ping 192.168.2.12 -t　 ping 192.168.2.11 -t　 ping 192.168.2.101 -t

(5) 按要求配置,划分 VLAN 及端口并进行验证。观察 ping 窗口的连通变化。

VLAN 10 (5~8)　VLAN 20 (9~12)　VLAN 30 (13~16)

(6) 交换机级联端口 A1、A4 和 B4、B7 设置为 Trunk,并进行验证。观察 ping 窗口的连通变化。

(7) 聚合 A1、A4 和 B4、B7,新端口形成,并进行验证。观察 ping 窗口的连通变化。

(8) 观察 PC1 ping PC2 通,实验成功。

(9) 按照作业内容,提取 A、B2 两台设备配置,撰写实验报告。

实验十二　认识三层交换机

一、实验目的

1. 熟悉高端三层交换机的外观。
2. 了解高端三层交换机各端口的名称和作用。
3. 学会使用 TFTP 服务器对三层交换机版本进行升级和老版本的备份。
4. 学会使用相关命令对三层交换机的接口地址进行配置。

二、相关知识

交换机的分类方法有很多种，按照不同的原则，交换机可以分成各种不同的类别：按照网络 OSI 七层模型来划分，可以将交换机划分为二层交换机、三层交换机、多层交换机。二层交换机是按照 MAC 地址进行数据帧的过滤和转发，这种交换机是目前最常见的交换机。三层交换机采用"一次路由，多次交换"的原理，基于 IP 地址转发数据包。部分三层交换机也具有四层交换机的一些功能，例如依据端口号进行转发。四层交换机以及四层以上的交换机都可以称为内容型交换机，一般使用在大型的网络数据中心。按照外观和架构的特点，可以将局域网交换机划分为机箱式交换机、机架式交换机、桌面式交换机。机箱式交换机外观比较庞大，这种交换机所有的部件都是可插拔的部件(一般称之为模块)，灵活性非常好。在实际的组网中，可以根据网络的要求选择不同的模块。模块可以分为几大类：一类是管理模块，它相当于计算机的主板和 CPU，用于管理整个交换机的工作；一类是应用模块，相当于计算机的 I/O 模块，负责连接其他网络设备和网络终端；另外还有电源模块、风扇模块等。在购买机箱式交换机时，需要分别购买机箱、管理模块、应用模块以及电源模块。机箱式交换机一般都是三层交换机或者多层交换机。在网络设计中，由于机箱式交换机性能和稳定性都比较卓越，因此价格比较昂贵，一般定位在核心层交换机或者汇聚层交换机。

三、实验设备

1. DCRS-7604(或 6804)交换机 1 台。
2. PC 机 1 台。
3. 交换机 Console 线 1 根。

四、实验拓扑

将 PC 的串口和交换机的 Console 口用 Console 线连接，如图 1-34 所示。

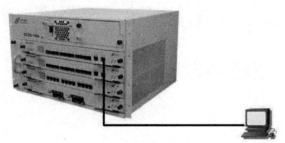

图 1-34　用 Console 线连接 PC 和交换机

五、实验要求

1. 正确认识交换机上各模块和物理端口名称。
2. 熟悉每个物理端口在配置界面中的对应名称。

六、实验步骤

第一步：认识交换机的端口，如图 1-35 所示。

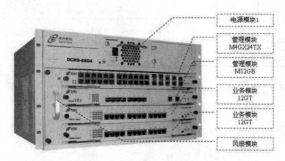

图 1-35　交换机端口示意图

第二步：以 MRS-7604-M12GB 为例(如图 1-36 所示)，交换机的模块 MRS-7604-M12GB 是 DCRS-7604 交换机的主控交换模块，承担着系统状态的控制、路由的管理、用户接入的控制和管理，以及网络的维护。主控板插在单板插框的第 1、2 槽位中，支持主备冗余，实现热备份。同时它还带有 12 个 SFP 形式的千兆光纤接口。

交换机前面板 MRS-7604-M12GB 提供了 12 个 SFP 端口，同时还提供了 1 个 Console 配置口(即控制台)，以及 1 个 10/100Base-T Ethernet 口(即管理网口)。

图 1-36　MRS-7604-M12GB 前面板示意图

1. MRS-7604-M12GB 前面板指示灯的状态说明如表 1-17 所示。

表 1-17　MRS-7604-M12GB 指示灯说明

LED 指示灯	面板标记	状态	含义
电源指示灯	PWR	亮(绿色)	板卡加电
		灭	板卡断电
运行指示灯	RUN	亮(绿色 1HZ 闪烁)	板卡运行状态正常
		亮(绿色 8HZ 闪烁)	系统加载中
		亮(黄色 8HZ 闪烁)	系统关闭中
		亮(红色 8HZ 闪烁)	运行状态故障
		灭	板卡已关闭，可拔出
主备指示灯	M/S	亮(绿色)	主用
		灭	备用
风扇指示灯	FAN	亮(绿色)	风扇在位
		灭	风扇不在位
SFP 接口指示灯			

(续表)

LED 指示灯	面板标记	状态	含义
状态指示灯	Link	亮(绿色)	SFP 收发器网络连接正常
		灭	SFP 收发器没有网络连接
传输指示灯	Act	闪(绿)	正在发送或接收数据

MRS-7604-M12GB 提供了 12 个 SFP(Mini GBIC)千兆光纤收发器插槽。

MRS-7604-M12GB 支持以下几种类型的 SFP 收发器。

(1) SFP-SX 收发器。

(2) SFP-LX 10 公里收发器。

(3) SFP-LH-40 40 公里中距收发器。

(4) SFP-LH-70 70 公里长距收发器。

(5) SFP-LH-120 120 公里超长距收发器。

各 SFP 收发器传输距离如表 1-18 所示。

表 1-18　SFP 收发器传输距离说明

接口形式	规格
SFP-SX 收发器	62.5/125um 多模光纤：275 米
	50/125um 多模光纤：550 米
SFP-LX 收发器	9/125um 单模光纤：10 公里
SFP-LH-40 收发器	9/125um 单模光纤：40 公里
SFP-LH-70 收发器	9/125um 单模光纤：70 公里
SFP-LH-120 收发器	9/125um 单模光纤：120 公里

2. 前面板控制台。MRS-7604-M12GB 提供了一个 RJ-45(母)Console 串口，如表 1-19 所示。通过该控制台，用户可以连接后台终端计算机，进行系统的调试、配置、维护、管理以及主机软件程序的加载等工作。

表 1-19　MRS-7604-M12GB 控制台说明

属性	规格
接头	RJ-45(母)
接口标准	RS-232
波特率	9600bps(默认)
支持服务	与字符终端连接
	与 PC 串口连接，并在 PC 上运行终端仿真程序

3. 前面板管理网口。MRS-7604-M12GB 提供了一个 RJ-45(母)Ethernet 接口，如表 1-20 所示，通过该网口，用户可以连接后台 PC，进行程序加载工作；也可通过该网口连接远端的网管工作站等设备，实现远程管理。

表 1-20　MRS-7604-M12GB 管理网口说明

属性	规格
接头	RJ-45(母)
接口标准	10/100Mbps 自适应
	5 类非屏蔽双绞线(UTP)：300 米

4. 前面板复位按键。MRS-7604-M12GB 提供了一个 RESET 复位按键，用于复位单板。前面板热拔按键(SWAP)MRS-7604-M12GB 提供了一个 SWAP 热拔按键，用于在设备运行过程中热插拔此模块。当用户准备将模块拔下时应先按下 SWAP 按键。此时模块做热插拔的准备工作，同时将系统运行指示灯(RUN)置为黄色 8Hz 闪烁。当 RUN 灯灭后，表明板卡已关闭，可以热插拔。

第三步：进入交换机配置界面。

```
DCRS-7604#show run
Current configuration:
!
     hostname DCRS-7604
!
Vlan 1
   vlan 1
!
Interface Ethernet1/1
!
Interface Ethernet1/2
……
Interface Ethernet2/27
!
Interface Ethernet2/28
!
Interface Ethernet0
!
DCRS-7604#
```

第四步：了解高端三层交换机各端口的名称。

Interface Ethernet1/1	表示第一模块的第 1 个接口。
Interface Ethernet1/28	表示第一模块的第 28 个接口。
Interface Ethernet2/1	表示第二模块的第 1 个接口。
Interface Ethernet2/28	表示第二模块的第 28 个接口。
Interface Ethernet0	表示 1 个 10/100Base-T Ethernet 口，即管理网口。

七、注意事项

DCRS-7604/6804 等调试界面类似于 DCS-3926S，基本的命令格式都一样。

八、课后练习

1. 熟悉交换机的各种配置模式。
2. 掌握交换机 CLI 界面的调试技巧。
3. 了解交换机恢复出厂设置及其基本配置方法。
4. 使用 Telnet 方式管理交换机。
5. 使用 Web 界面管理交换机。
6. 执行交换机文件备份。
7. 进行交换机系统升级和文件还原。
8. 解决密码丢失的问题。
9. 在 Bootrom 下进行升级配置。

实验十三　多层交换机 VLAN 的划分和 VLAN 间路由

一、实验目的

1. 掌握 VLAN 原理。
2. 学会使用各种多层交换设备进行 VLAN 的划分。
3. 理解 VLAN 之间路由的原理和实现方法。

二、相关知识

多层交换机具备第二层交换(基于 MAC 地址)和第三层交换(基于 IP 地址)的能力，可以实现 VLAN 间的路由，同时保持高效的交换速度。通过在多层交换机上实施 VLAN 技术，可以进行 VLAN 的划分和 VLAN 间路由配置从而实现不同 VLAN 之间的通信。

多层交换机 VLAN 的划分和 VLAN 间路由的实验场景如下。

(1) 校园网络：校园网络中的不同部门可能需要访问不同的资源，VLAN 划分和 VLAN 间路由可以帮助实现资源的隔离和访问控制。例如，软件实验室的 IP 地址段是 192.168.10.0/24，多媒体实验室的 IP 地址段是 192.168.20.0/24。为了保证它们之间的数据互不干扰，也不影响各自的通信效率，我们划分了 VLAN，使两个实验室属于不同的 VLAN。尽管如此，两个实验室有时也需要相互通信，此时就要利用三层交换机划分 VLAN。

(2) 企业网络管理：在企业网络中，通过 VLAN 划分和 VLAN 间路由可以提高网络的安全性和可管理性。

(3) 数据中心网络：数据中心中的服务器和存储设备可能分布在不同的 VLAN 中，VLAN 间路由可以实现这些设备之间的通信。

三、实验设备

1. DCRS-7604(或 6804 和 5526S)交换机 1 台。
2. PC 机 2 台。
3. Console 线 1 根。
4. 直通网线若干。

四、实验拓扑

使用一台交换机和两台 PC,将其中 PC2 作为控制台终端,使用 Console 口配置方式。使用两根网线分别将 PC1 和 PC2 连接到交换机的 RJ-45 接口上。该实验拓扑结构如图 1-37 所示。

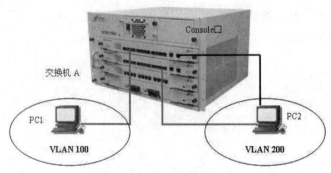

图 1-37 实验拓扑图

五、实验要求

在交换机上划分两个基于端口的 VLAN：VLAN 100 和 VLAN 200。配置后,使 VLAN 100 的成员可以互相访问,VLAN 200 的成员也可以互相访问,但 VLAN 100 和 VLAN 200 的成员之间不能互相访问。具体配置如表 1-21 所示。

表 1-21 VLAN 划分表

VLAN	端口成员
100	1/1~1/12
200	1/13~1/24

PC1 和 PC2 的网络设置如表 1-22 所示。

表 1-22 PC1 和 PC2 配置表

设备	端口	IP1	网关 1	IP2	网关 2	Mask
交换机 A		192.168.2.1	无	192.168.2.1	无	255.255.255.0
VLAN 100		无	无	192.168.10.1	无	255.255.255.0
VLAN 200		无	无	192.168.20.1	无	255.255.255.0

(续表)

设备	端口	IP1	网关1	IP2	网关2	Mask
PC1	1~12	192.168.2.101	无	192.168.10.101	192.168.10.1	255.255.255.0
PC2	13~24	192.168.2.102	无	192.168.20.102	192.168.20.1	255.255.255.0

各设备的 IP 地址首先按照 IP1 配置，使用 PC1 ping PC2，应该不通；再按照 IP2 配置地址，并在交换机上配置 VLAN 接口 IP 地址，使用 PC1 ping PC2，发现能通，该通信属于 VLAN 间通信，要经过三层设备的路由。

若实验结果和理论相符，则本实验完成。

六、实验步骤

神州数码设备的配置步骤如下。

第一步：交换机恢复出厂设置。

```
switch#set default
switch#write
switch#reload
```

第二步：为交换机配置管理 IP 地址。

```
switch#config
switch(Config)#interface vlan 1
switch(Config-If-Vlan1)#ip address 192.168.2.1 255.255.255.0
switch(Config-If-Vlan1)#no shutdown
switch(Config-If-Vlan1)#exit
switch(Config)#exit
```

第三步：创建 VLAN 100 和 VLAN 200。

```
switch(Config)#
switch(Config)#vlan 100
switch(Config-Vlan100)#exit
switch(Config)#vlan 200
switch(Config-Vlan200)#exit
switch(Config)#
```

验证配置：

```
switch#show vlan
VLAN Name            Type     Media    Ports
---- -------------   -------  -------  -----------------------------------
1    default         Static   ENET     Ethernet1/1    Ethernet1/2
                                       Ethernet1/3    Ethernet1/4
                                       Ethernet1/5    Ethernet1/6
                                       Ethernet1/7    Ethernet1/8
```

				Ethernet1/9	Ethernet1/10
				Ethernet1/11	Ethernet1/12
				Ethernet1/13	Ethernet1/14
				Ethernet1/15	Ethernet1/16
				Ethernet1/17	Ethernet1/18
				Ethernet1/19	Ethernet1/20
				Ethernet1/21	Ethernet1/22
				Ethernet1/23	Ethernet1/24
				Ethernet1/25	Ethernet1/26
				Ethernet1/27	Ethernet1/28
100	VLAN0100	Static	ENET		
200	VLAN0200	Static	ENET		

第四步：给 VLAN 100 和 VLAN 200 添加端口。

```
switch(Config)#vlan 100                              ！进入 VLAN 100
switch(Config-Vlan100)#switchport interface ethernet 1/1-12
Set the port Ethernet1/1 access vlan 100 successfully
Set the port Ethernet1/2 access vlan 100 successfully
Set the port Ethernet1/3 access vlan 100 successfully
Set the port Ethernet1/4 access vlan 100 successfully
Set the port Ethernet1/5 access vlan 100 successfully
Set the port Ethernet1/6 access vlan 100 successfully
Set the port Ethernet1/7 access vlan 100 successfully
Set the port Ethernet1/8 access vlan 100 successfully
Set the port Ethernet1/9 access vlan 100 successfully
Set the port Ethernet1/10 access vlan 100 successfully
Set the port Ethernet1/11 access vlan 100 successfully
Set the port Ethernet1/12 access vlan 100 successfully
switch(Config-Vlan100)#exit
switch(Config)#vlan 200                              ！进入 VLAN 200
switch(Config-Vlan200)#switchport interface ethernet 1/13-24
Set the port Ethernet1/13 access vlan 200 successfully
Set the port Ethernet1/14 access vlan 200 successfully
Set the port Ethernet1/15 access vlan 200 successfully
Set the port Ethernet1/16 access vlan 200 successfully
Set the port Ethernet1/17 access vlan 200 successfully
Set the port Ethernet1/18 access vlan 200 successfully
Set the port Ethernet1/19 access vlan 200 successfully
Set the port Ethernet1/20 access vlan 200 successfully
Set the port Ethernet1/21 access vlan 200 successfully
Set the port Ethernet1/22 access vlan 200 successfully
Set the port Ethernet1/23 access vlan 200 successfully
Set the port Ethernet1/24 access vlan 200 successfully
switch(Config-Vlan200)#exit
```

验证配置：

```
switch#show vlan
VLAN Name              Type    Media   Ports
---- ---------------   ------  ------  ------------------------------------
1    default           Static  ENET    Ethernet1/25    Ethernet1/26
                                       Ethernet1/27    Ethernet1/28
100  VLAN0100          Static  ENET    Ethernet1/1     Ethernet1/2
                                       Ethernet1/3     Ethernet1/4
                                       Ethernet1/5     Ethernet1/6
                                       Ethernet1/7     Ethernet1/8
                                       Ethernet1/9     Ethernet1/10
                                       Ethernet1/11    Ethernet1/12
200  VLAN0200          Static  ENET    Ethernet1/13    Ethernet1/14
                                       Ethernet1/15    Ethernet1/16
                                       Ethernet1/17    Ethernet1/18
                                       Ethernet1/19    Ethernet1/20
                                       Ethernet1/21    Ethernet1/22
                                       Ethernet1/23    Ethernet1/24
switch#
```

第五步：验证实验。结果如表 1-23 所示。

表 1-23 验证结果

PC1 的位置	PC2 的位置	动作	结果
1/1-1/12 端口		1/13-1/24 端口 PC1 ping	不通

第六步：添加 VLAN 地址。

```
switch(Config)#interface vlan 100
switch(Config-If-Vlan100)#%Feb 13   15:47:45 2006 %LINK-5-CHANGED:   Interface
Vlan100, changed state to UP%Feb 13 15:47:45 2006 %LINEPROTO-5-UPDOWN: Line
protocol
on Interface Vlan100, changed state to UP
switch(Config-If-Vlan100)#
switch(Config-If-Vlan100)#ip address 192.168.10.1 255.255.255.0
switch(Config-If-Vlan100)#no shut
switch(Config-If-Vlan100)#exit
switch(Config)#interface vlan 200
switch(Config-If-Vlan200)#%Feb 13   15:48:06 2006 %LINK-5-CHANGED:   Interface
Vlan200, changed   state to UPswitch(Config-If-Vlan200)#ip address 192.168.20.1
255.255.255.0
switch(Config-If-Vlan200)#no shut
switch(Config-If-Vlan200)#exit
switch(Config)#
```

验证配置:

```
switch#show ip route
Total route items is 3, the matched route items is 3
Codes: C - connected, S - static, R - RIP derived, O - OSPF derived
       A - OSPF ASE, B - BGP derived, D - DVMRP derived
Destination      Mask            Nexthop        Interface       Preference
C   192.168.2.0   255.255.255.0   0.0.0.0        Vlan1           0
C   192.168.10.0  255.255.255.0   0.0.0.0        Vlan100         0
C   192.168.20.0  255.255.255.0   0.0.0.0        Vlan200         0
switch#
```

思科设备的配置命令如下:

```
Switch>enable
Switch#configure terminal
Enter configuration commands, one per line.   End with CNTL/Z.
Switch(config)#hostname RS-A
RS-A(config)#interface vlan 1
RS-A(config-if)#ip address 92.168.2.1 255.255.255.0
RS-A(config-if)#no shutdown
RS-A(config-if)#
%LINK-5-CHANGED: Interface Vlan1, changed state to up
RS-A(config-if)#exit
RS-A(config)#vl
RS-A(config)#vlan 100
RS-A(config-vlan)#exit
RS-A(config)#vlan 200
RS-A(config-vlan)#exit
RS-A(config)#interface range fastEthernet 0/1-12
RS-A(config-if-range)#switchport access vlan 100
RS-A(config-if-range)#ex
RS-A(config)#interface range fastEthernet 0/13-24
RS-A(config-if-range)#switchport access vlan 200
RS-A(config-if-range)#exit
RS-A(config)#interface vlan 100
RS-A(config-if)#
%LINK-5-CHANGED: Interface Vlan100, changed state to up
RS-A(config-if)#ip address 192.168.10.1 255.255.255.0
RS-A(config-if)#no shutdown
RS-A(config-if)#ex
RS-A(config)#interface vlan 200
RS-A(config-if)#
%LINK-5-CHANGED: Interface Vlan200, changed state to up
RS-A(config-if)#ip address 192.168.20.1 255.255.255.0
RS-A(config-if)#no shutdown
```

```
RS-A(config-if)#ex
RS-A(config)#^Z
RS-A#
%SYS-5-CONFIG_I: Configured from console by console
RS-A#sh vl
RS-A#sh vlan
```

VLAN	Name	Status	Ports
1	default	active	Gig0/1, Gig0/2
100	VLAN0100	active	Fa0/1, Fa0/2, Fa0/3, Fa0/4
			Fa0/5, Fa0/6, Fa0/7, Fa0/8
			Fa0/9, Fa0/10, Fa0/11, Fa0/12
200	VLAN0200	active	Fa0/13, Fa0/14, Fa0/15, Fa0/16
			Fa0/17, Fa0/18, Fa0/19, Fa0/20
			Fa0/21, Fa0/22, Fa0/23, Fa0/24
1002	fddi-default		act/unsup
1003	token-ring-default		act/unsup
1004	fddinet-default		act/unsup
1005	trnet-default		act/unsup

VLAN	Type	SAID	MTU	Parent	RingNo	BridgeNo	Stp	BrdgMode	Trans1	Trans2
1	enet	100001	1500	-	-	-	-	-	0	0
100	enet	100100	1500	-	-	-	-	-	0	0
200	enet	100200	1500	-	-	-	-	-	0	0
1002	fddi	101002	1500	-	-	-	-	-	0	0
1003	tr	101003	1500	-	-	-	-	-	0	0
1004	fdnet	101004	1500	-	-	-	ieee	-	0	0
1005	trnet	101005	1500	-	-	-	ibm	-	0	0

Remote SPAN VLANs

Primary Secondary Type   Ports
------- --------- --------------- --

```
RS-A#sh run
Building configuration...
Current configuration : 1843 bytes
!
version 12.2
no service timestamps log datetime msec
no service timestamps debug datetime msec
no service password-encryption
!
hostname RS-A
!
!
!
!
```

!
spanning-tree mode pvst
!
interface FastEthernet0/1
　switchport access vlan 100
!
interface FastEthernet0/2
　switchport access vlan 100
!
interface FastEthernet0/3
　switchport access vlan 100
!
interface FastEthernet0/4
　switchport access vlan 100
!
interface FastEthernet0/5
　switchport access vlan 100
!
interface FastEthernet0/6
　switchport access vlan 100
!
interface FastEthernet0/7
　switchport access vlan 100
!
interface FastEthernet0/8
　switchport access vlan 100
!
interface FastEthernet0/9
　switchport access vlan 100
!
interface FastEthernet0/10
　switchport access vlan 100
!
interface FastEthernet0/11
　switchport access vlan 100
!
interface FastEthernet0/12
　switchport access vlan 100
!
interface FastEthernet0/13
　switchport access vlan 200
!
interface FastEthernet0/14
　switchport access vlan 200
!
interface FastEthernet0/15

```
  switchport access vlan 200
!
interface FastEthernet0/16
  switchport access vlan 200
!
interface FastEthernet0/17
  switchport access vlan 200
!
interface FastEthernet0/18
  switchport access vlan 200
!
interface FastEthernet0/19
  switchport access vlan 200
!
interface FastEthernet0/20
  switchport access vlan 200
!
interface FastEthernet0/21
  switchport access vlan 200
!
interface FastEthernet0/22
  switchport access vlan 200
!
interface FastEthernet0/23
  switchport access vlan 200
!
interface FastEthernet0/24
  switchport access vlan 200
!
interface GigabitEthernet0/1
!
interface GigabitEthernet0/2
!
interface Vlan1
  ip address 92.168.2.1 255.255.255.0
!
interface Vlan100
  ip address 192.168.10.1 255.255.255.0
!
interface Vlan200
  ip address 192.168.20.1 255.255.255.0
!
!
!
!
line con 0
```

```
!
line vty 0 4
  login
line vty 5 15
  login
!
!
end
RS-A#
```

第七步：验证实验。结果如表 1-24 所示。

表 1-24 验证结果

PC1 的位置	PC2 的位置	动作	结果
1/1-1/12 端口	1/13-1/24 端口	PC1 ping PC2	通

七、注意事项和排错

三层交换机可以在多个 VLAN 接口上配置 IP 地址。

八、配置序列

```
switch#show run
Current configuration:
!
    hostname switch
!
Vlan 1
    vlan 1
!
Vlan 100
    vlan 100
!
Vlan 200
    vlan 200
!
Interface Ethernet1/1
    switchport access vlan 100
!
Interface Ethernet1/2
    switchport access vlan 100
!
Interface Ethernet1/3
    switchport access vlan 100
!
```

```
Interface Ethernet1/4
    switchport access vlan 100
!
Interface Ethernet1/5
    switchport access vlan 100
!
Interface Ethernet1/6
    switchport access vlan 100
!
Interface Ethernet1/7
    switchport access vlan 100
!
Interface Ethernet1/8
    switchport access vlan 100
!
Interface Ethernet1/9
    switchport access vlan 100
!
Interface Ethernet1/10
    switchport access vlan 100
!
Interface Ethernet1/11
    switchport access vlan 100
!
Interface Ethernet1/12
    switchport access vlan 100
!
Interface Ethernet1/13
    switchport access vlan 200
!
Interface Ethernet1/14
    switchport access vlan 200
!
Interface Ethernet1/15
    switchport access vlan 200
!
Interface Ethernet1/16
    switchport access vlan 200
!
Interface Ethernet1/17
    switchport access vlan 200
!
Interface Ethernet1/18
    switchport access vlan 200
!
Interface Ethernet1/19
```

```
    switchport access vlan 200
!
Interface Ethernet1/20
    switchport access vlan 200
!
Interface Ethernet1/21
    switchport access vlan 200
!
Interface Ethernet1/22
    switchport access vlan 200
!
Interface Ethernet1/23
    switchport access vlan 200
!
Interface Ethernet1/24
    switchport access vlan 200
!
Interface Ethernet1/25
!
Interface Ethernet1/26
!
Interface Ethernet1/27
!
Interface Ethernet1/28
!
!
interface Vlan1
    interface vlan 1
    ip address 192.168.2.1 255.255.255.0
!
interface Vlan100
    interface vlan 100
    ip address 192.168.10.1 255.255.255.0
!
interface Vlan200
    interface vlan 200
    ip address 192.168.20.1 255.255.255.0
!
Interface Ethernet0
switch#
```

九、思考题

第二次配置 IP 地址时，如果没有为 PC 配置网关，它还会通信吗？为什么？

十、课后练习

场景描述：假设你是一家大型企业的网络管理员，公司有三个主要办公地点，分别位于不同的建筑中。为了优化网络管理和提高安全性，公司决定实施 VLAN 策略。你需要确保以下部门在所有办公地点之间能够实现跨交换机通信：

- 市场部(VLAN 10)
- 技术部(VLAN 20)
- 财务部(VLAN 30)

1. VLAN 规划与设计：设计一个网络拓扑图，展示三个办公地点的交换机连接方式，包括交换机之间的 Trunk 链路。

2. 多层交换机配置：
(1) 在三个地点的多层交换机上创建上述 VLAN。
(2) 将特定端口分配给对应的 VLAN。
(3) 配置交换机之间的 Trunk 链路。
(4) 配置 VLAN 间路由。

3. 验证配置：
(1) 使用 show vlan brief、show interfaces trunk 和 show ip route 命令验证配置。
(2) 通过在不同 VLAN 的计算机之间进行通信测试，验证 VLAN 间的隔离和通信。

实验十四　多层交换机实现二层交换机 VLAN 间路由

一、实验目的

1. 理解多层交换机的路由原理。
2. 了解多层交换机在实际网络中的常用配置。
3. 回顾二层交换机 VLAN 的划分方法。
4. 进一步理解 IEEE802.1Q 的原理和使用方法。

二、相关知识

多层交换机实现二层交换机 VLAN 间路由的实验旨在帮助学生理解如何在多层交换机上配置 VLAN 间路由，以实现不同 VLAN 之间的通信，同时保持二层交换的高效性。二层交换机通常作为接入层交换机，根据连接用户的不同划分为多个 VLAN。在一些情况下，同一个 VLAN 可能分布在不同的交换机上。这些二层交换机通过一台三层交换机进行汇聚。

三、实验设备

1. DCRS-7604(或 6804 或 5526S)交换机 1 台。

2. DCS-3926S 交换机 1~2 台。
3. PC 机 2~4 台。
4. Console 线 1~3 根。
5. 直通网线若干。

四、实验拓扑

图 1-38 所示为多层交换机实现二层交换机 VLAN 之间路由的实验拓扑图。

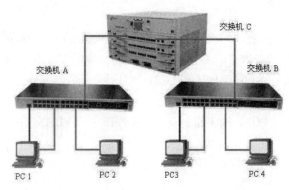

图 1-38 实验拓扑图

五、实验要求

在交换机 A 和交换机 B 上分别划分两个基于端口的 VLAN：VLAN 100 和 VLAN 200。具体配置如表 1-25 所示。

表 1-25 交换机 A 和交换机 B 的 VLAN 划分

VLAN	端口成员
100	0/0/1~0/0/8
200	0/0/9~0/0/16
Trunk 口	24

在交换机 C 上也划分两个基于端口的 VLAN：VLAN 100 和 VLAN 200。把端口 1 和端口 2 都设置成 Trunk 口。具体配置如表 1-26 所示。

表 1-26 VLAN 的配置

VLAN	IP	Mask
100	192.168.10.1	255.255.255.0
200	192.168.20.1	255.255.255.0
Trunk 口		0/0/1 和 0/0/2

交换机 A 的 24 口连接交换机 C 的 1 口，交换机 B 的 24 口连接交换机 C 的 2 口。具

体配置如表 1-27 所示。

表 1-27 PC 的配置

设备	IP 地址	gateway	Mask
PC1	192.168.10.11	192.168.10.1	255.255.255.0
PC2	192.168.20.22	192.168.20.1	255.255.255.0
PC3	192.168.10.33	192.168.10.1	255.255.255.0

验证实验结果。

1. 不给 PC 设置网关：PC1、PC3 分别接在不同交换机 VLAN 100 的成员端口 1~8 上，两台 PC 互相可以 ping 通；PC2、PC4 分别接在不同交换机 VLAN 的成员端口 9~16 上，两台 PC 互相可以 ping 通；PC1、PC3 和 PC2、PC4 接在不同 VLAN 的成员端口上则互相 ping 不通。

2. 给 PC 设置网关：PC1、PC3 和 PC2、PC4 接在不同 VLAN 的成员端口上也可以互相 ping 通。

若实验结果和理论相符，则本实验完成。

六、实验步骤

神州数码设备的配置步骤如下。

第一步：交换机恢复出厂设置。

```
switch#set default
switch#write
switch#reload
```

第二步：给交换机设置标示符和管理 IP。

交换机 A：

```
switch(Config)#hostname switchA
switchA(Config)#interface vlan 1
switchA(Config-If-Vlan1)#ip address 192.168.2.11 255.255.255.0
switchA(Config-If-Vlan1)#no shutdown
switchA(Config-If-Vlan1)#exit
switchA(Config)#
```

交换机 B：

```
switch(Config)#hostname switchB
switchB(Config)#interface vlan 1
switchB(Config-If-Vlan1)#ip address 192.168.2.12 255.255.255.0
switchB(Config-If-Vlan1)#no shutdown
switchB(Config-If-Vlan1)#exit
```

switchB(Config)#

交换机 C：

DCRS-7604#config
DCRS-7604(Config)#
DCRS-7604(Config)#hostname switchC
switchC(Config)#interface vlan 1
switchC(Config-If-Vlan1)#ip address 192.168.2.13 255.255.255.0
switchC(Config-If-Vlan1)#no shutdown
switchC(Config-If-Vlan1)#exit
switchC(Config)#exit
switchC#

第三步：在交换机中创建 VLAN 100 和 VLAN 200，并添加端口。

交换机 A：

switchA(Config)#vlan 100
switchA(Config-Vlan100)#
switchA(Config-Vlan100)#switchport interface ethernet 0/0/1-8
switchA(Config-Vlan100)#exit
switchA(Config)#vlan 200
switchA(Config-Vlan200)#switchport interface ethernet 0/0/9-16
switchA(Config-Vlan200)#exit
switchA(Config)#

验证配置：

switchA#show vlan

VLAN	Name	Type	Media	Ports	
1	default	Static	ENET	Ethernet0/0/17	Ethernet0/0/18
				Ethernet0/0/19	Ethernet0/0/20
				Ethernet0/0/21	Ethernet0/0/22
				Ethernet0/0/23	Ethernet0/0/24
100	VLAN0100	Static	ENET	Ethernet0/0/1	Ethernet0/0/2
				Ethernet0/0/3	Ethernet0/0/4
				Ethernet0/0/5	Ethernet0/0/6
				Ethernet0/0/7	Ethernet0/0/8
200	VLAN0200	Static	ENET	Ethernet0/0/9	Ethernet0/0/10
				Ethernet0/0/11	Ethernet0/0/12
				Ethernet0/0/13	Ethernet0/0/14
				Ethernet0/0/15	Ethernet0/0/16

switchA#

交换机 B：配置与交换机 A 一样。
第四步：设置交换机 Trunk 端口。
交换机 A：

```
switchA(Config)#interface ethernet 0/0/24
switchA(Config-Ethernet0/0/24)#switchport mode trunk
Set the port Ethernet0/0/24 mode TRUNK successfully
switchA(Config-Ethernet0/0/24)#switchport trunk allowed vlan all
set the port Ethernet0/0/24 allowed vlan successfully
switchA(Config-Ethernet0/0/24)#exit
switchA(Config)#
```

验证配置：

```
switchA#show vlan
VLAN  Name       Type    Media   Ports
1     default    Static  ENET    Ethernet0/0/17      Ethernet0/0/18
                                 Ethernet0/0/19      Ethernet0/0/20
                                 Ethernet0/0/21      Ethernet0/0/22
                                 Ethernet0/0/23
                                 Ethernet0/0/24(T)
100   VLAN0100   Static  ENET    Ethernet0/0/1       Ethernet0/0/2
                                 Ethernet0/0/3       Ethernet0/0/4
                                 Ethernet0/0/5       Ethernet0/0/6
                                 Ethernet0/0/7       Ethernet0/0/8
                                 Ethernet0/0/24(T)
200   VLAN0200   Static  ENET    Ethernet0/0/9       Ethernet0/0/10
                                 Ethernet0/0/11      Ethernet0/0/12
                                 Ethernet0/0/13      Ethernet0/0/14
                                 Ethernet0/0/15      Ethernet0/0/16
                                 Ethernet0/0/24(T)
switchA#
```

24 口已经出现在 VLAN 1、VLAN 100 和 VLAN 200 中，并且 24 口不是一个普通端口，是 Tagged 端口。

交换机 B：配置与交换机 A 一样。
交换机 C：

```
switchC(Config)#vlan 100
switchC(Config-Vlan100)#exit
switchC(Config)#vlan 200
switchC(Config-Vlan200)#exit
switchC(Config)#interface ethernet 0/0/1-2
switchC(Config-Port-Range)#switchport mode trunk
```

```
Set the port Ethernet 0/0/1 mode TRUNK successfully
Set the port Ethernet 0/0/2 mode TRUNK successfully
switchC(Config-Port-Range)#switchport trunk allowed vlan all
set the port Ethernet 0/0/1 allowed vlan successfully
set the port Ethernet 0/0/2 allowed vlan successfully
switchC(Config-Port-Range)#exit
switchC(Config)#exit
```

验证配置：

```
switchC#show vlan
```

VLAN	Name	Type	Media	Ports	
1	default	Static	ENET	Ethernet1/1(T)	Ethernet1/2(T)
				Ethernet1/3	Ethernet1/4
				Ethernet1/5	Ethernet1/6
				Ethernet1/7	Ethernet1/8
				Ethernet1/9	Ethernet1/10
				Ethernet1/11	Ethernet1/12
				Ethernet1/13	Ethernet1/14
				Ethernet1/15	Ethernet1/16
				Ethernet1/17	Ethernet1/18
				Ethernet1/19	Ethernet1/20
				Ethernet1/21	Ethernet1/22
				Ethernet1/23	Ethernet1/24
				Ethernet1/25	Ethernet1/26
				Ethernet1/27	Ethernet1/28
100	VLAN0100	Static	ENET	Ethernet1/1(T)	Ethernet1/2(T)
200	VLAN0200	Static	ENET	Ethernet1/1(T)	Ethernet1/2(T)

```
switchC#
```

第五步：交换机 C 添加 VLAN 地址。

```
switchC(Config)#interface vlan 100
switchC(Config-If-Vlan100)#ip address 192.168.10.1 255.255.255.0
switchC(Config-If-Vlan100)#no shut
switchC(Config-If-Vlan100)#exit
switchC(Config)#interface vlan 200
switchC(Config-If-Vlan200)#ip address 192.168.20.1 255.255.255.0
switchC(Config-If-Vlan200)#no shutdown
switchC(Config-If-Vlan200)#exit
switchC(Config)#
```

验证配置：

```
switch#show ip route
```

```
Total route items is 3, the matched route items is 3
Codes: C - connected, S - static, R - RIP derived, O - OSPF derived
       A - OSPF ASE, B - BGP derived, D - DVMRP derived
Destination        Mask              Nexthop        Interface      Preference
C   192.168.2.0    255.255.255.0     0.0.0.0        Vlan1          0
C   192.168.10.0   255.255.255.0     0.0.0.0        Vlan100        0
C   192.168.20.0   255.255.255.0     0.0.0.0        Vlan200        0
switch#
```

思科设备配置命令如下。

二层交换机 A：

```
Switch(config)#hostname S-A
S-A(config)#vlan 100
S-A(config-vlan)#ex
S-A(config)#vlan 200
S-A(config-vlan)#ex
S-A(config)#interface range fastEthernet 0/1-8
S-A(config-if-range)#switchport access vlan 100
S-A(config-if-range)#ex
S-A(config)#interface range fastEthernet 0/9-16
S-A(config-if-range)#switchport access vlan 200
S-A(config-if-range)#ex
S-A(config)#interface fastEthernet 0/24
S-A(config-if)#switchport mode trunk
S-A(config-if)#
```

二层交换机 B：配置与交换机 A 一样。

三层交换机 C：

```
Switch(config)#hostname RS-C
RS-C#
RS-C#configure
RS-C(config)#vlan 100
RS-C(config-vlan)#exit
RS-C(config)#vlan 200
RS-C(config-vlan)#exit
RS-C(config)#interface range fastEthernet 0/1-2
RS-C(config-if-range)#switchport trunk encapsulation dot1q
(先设置端口封装模式接入主干通道)
RS-C(config-if-range)#switchport mode trunk
RS-C(config-if-range)#exit
RS-C(config)#interface vlan 100
RS-C(config-if)#ip address 192.168.10.1 255.255.255.0
```

RS-C(config-if)#no shutdown
RS-C(config-if)#exit
RS-C(config)#interface vlan 200
RS-C(config-if)#ip address 192.168.20.1 255.255.255.0
RS-C(config-if)#no shutdown
RS-C(config-if)#exit
启动三层交换机的默认路由
RS-C(config)#ip routing
RS-C(config)#exit
RS-C#
RS-C#sh vlan

VLAN	Name	Status	Ports
1	default	active	Fa0/3, Fa0/4, Fa0/5, Fa0/6
			Fa0/7, Fa0/8, Fa0/9, Fa0/10
			Fa0/11, Fa0/12, Fa0/13, Fa0/14
			Fa0/15, Fa0/16, Fa0/17, Fa0/18
			Fa0/19, Fa0/20, Fa0/21, Fa0/22
			Fa0/23, Fa0/24, Gig0/1, Gig0/2
100	VLAN0100	active	
200	VLAN0200	active	
1002	fddi-default	act/unsup	
1003	token-ring-default	act/unsup	
1004	fddinet-default	act/unsup	
1005	trnet-default	act/unsup	

VLAN	Type	SAID	MTU	Parent	RingNo	BridgeNo	Stp	BrdgMode	Trans1	Trans2
1	enet	100001	1500	-	-	-	-	-	0	0
100	enet	100100	1500	-	-	-	-	-	0	0
200	enet	100200	1500	-	-	-	-	-	0	0
1002	fddi	101002	1500	-	-	-	-	-	0	0
1003	tr	101003	1500	-	-	-	-	-	0	0
1004	fdnet	101004	1500	-	-	-	ieee	-	0	0
1005	trnet	101005	1500	-	-	-	ibm	-	0	0

Remote SPAN VLANs

Primary Secondary Type   Ports
------- --------- ---------------- ------------------------------------

RS-C#sh run
Building configuration...
Current configuration : 1359 bytes
!

```
version 12.2
no service timestamps log datetime msec
no service timestamps debug datetime msec
no service password-encryption
!
hostname RS-C
!
!
spanning-tree mode pvst
!
!
!
!
interface FastEthernet0/1
  switchport trunk encapsulation dot1q
  switchport mode trunk
!
interface FastEthernet0/2
  switchport trunk encapsulation dot1q
  switchport mode trunk
!
interface FastEthernet0/3
!
interface FastEthernet0/4
.......
!
interface FastEthernet0/24
!
interface GigabitEthernet0/1
!
interface GigabitEthernet0/2
!
interface Vlan1
  no ip address
  shutdown
!
interface Vlan100
  ip address 192.168.10.1 255.255.255.0
!
interface Vlan200
  ip address 192.168.20.1 255.255.255.0
!
```

```
ip classless
!
ip flow-export version 9
!
!
RS-C#
RS-C#sh ip route
Codes: C - connected, S - static, I - IGRP, R - RIP, M - mobile, B - BGP
       D - EIGRP, EX - EIGRP external, O - OSPF, IA - OSPF inter area
       N1 - OSPF NSSA external type 1, N2 - OSPF NSSA external type 2
       E1 - OSPF external type 1, E2 - OSPF external type 2, E - EGP
       i - IS-IS, L1 - IS-IS level-1, L2 - IS-IS level-2, ia - IS-IS inter area
       * - candidate default, U - per-user static route, o - ODR
       P - periodic downloaded static route
Gateway of last resort is not set
C    192.168.10.0/24 is directly connected, Vlan100
C    192.168.20.0/24 is directly connected, Vlan200
```

第六步：验证实验。

1. PC 不配置网关，互相 ping，查看结果。
2. PC 配置网关，互相 ping，查看结果。

七、注意事项和排错

当执行 show ip route 命令时，如果某一个网段上没有活跃的设备连接到三层交换机，则该网段路由信息将不会显示在命令输出中。

八、思考题

如果两台三层交换机级联，如何配置 VLAN？是否需要将某些端口设置为 Trunk 模式？

实训项目四 三层交换机实现二层交换机不同 VLAN 间通信

一、实训目的

1. 利用 VLAN 间路由的基本概念和作用，实现三层交换机上不同 VLAN 之间的通信。
2. 在三层交换机上配置 VLAN 间路由，掌握 VLAN 路由的实现方式及相关配置命令。

二、实训设备

(1) DCS-3926S 交换机 2 台。

(2) PC 机 2 台。

(3) 直通网线 3 根。

三、实训拓扑

该实验拓扑结构如图 1-39 所示。

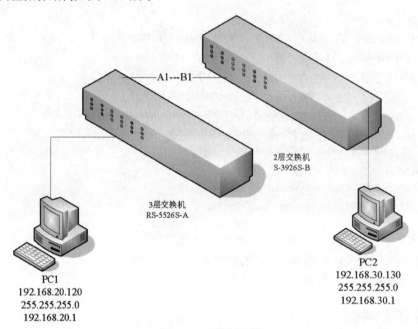

图 1-39　实验拓扑图

四、实训要求

参照如图 1-39 所示进行连线操作。

交换机 A 和交换机 B 做如下配置：

1. 划分 vl 10(5~8)、vl 20(9~12)、vl 30(13~16)。

2. 将交换机 A 和交换机 B 的级联端口设置为 Trunk 模式。

3. 设置 3 层交换机 RS-5526S-A；启用三层路由功能；设置 vl 10、vl 20、vl 30 的接口地址，具体如下。

- VLAN 10：int vl 10，192.168.10.1/24。
- VLAN 20：int vl 20，192.168.20.1/24。
- VLAN 30：int vl 30，192.168.30.1/24。

4. 根据连线位置正确配置 PC1、PC2 的地址(注意配上网关地址以及与 VLAN 的对应关系)。

实验结果：PC1---ping---PC2　通。

五、实训步骤

第一步：对交换机 A 和交换机 B 进行如下操作。
1. 恢复出厂设置(Set default、write、reload)。
2. 修改设备名称为：RS-A 和 S-B。

第二步：对 PC1 和 PC2 的本地连接 2 设置如下：

设备	IP 地址	gateway	Mask
PC1	192.168.20.120	192.168.20.1	255.255.255.0
PC2	192.168.30.130	192.168.30.1	255.255.255.0

注意：
将 PC1 和 PC2 的"本地连接"中的"默认网关"去掉。

第三步：在交换机 RS-A 上划分两个 VLAN，端口如表 1-28 所示。

表 1-28 交换机 RS-A 的 VLAN 划分

VLAN	端口成员
10	5~8
20	9~12
30	13~16
Trunk 口	0/0/1

第四步：在交换机 S-B 上也划分两个 VLAN，端口如表 1-29 所示。

表 1-29 交换机 S-B 的 VLAN 划分

VLAN	端口成员
10	5~8
20	9~12
30	13~16
Trunk 口	0/0/1

第五步：交换机 A 的 1 口连交换机 B 的 1 口。

根据 PC 网关与 VLAN 的对应关系，分析 PC 的连线位置。

PC1 网关 192.168.20.1 对应 A-VLAN 20 接口的 IP，所以 PC1 连接 A-VLAN 20 的 9~12 位置；

PC2 网关 192.168.30.1 对应 A-VLAN 30 接口的 IP，所以 PC1 连接 B-VLAN 30 的 13~16 位置。

第六步：在交换机 A 上进行如下操作。

```
RS-A#Config
RS-A (Config)#interface vlan 10
RS-A (Config-If-Vlan10)#ip   address 192.168.10.1   255.255.255.0
```

RS-A (Config-If-Vlan10)#no shutdown
RS-A (Config-If-Vlan10)#exit
RS-A (Config)#interface vlan 20
RS-A (Config-If-Vlan20)#ip   address   192.168.20.1   255.255.255.0
RS-A (Config-If-Vlan20)#no shutdown
RS-A (Config-If-Vlan20)#exit
RS-A (Config)#interface vlan 30
RS-A (Config-If-Vlan30)#ip   address   192.168.30.1   255.255.255.0
RS-A (Config-If-Vlan30)#no shutdown
RS-A (Config-If-Vlan30)#exit

验证配置：

RS-A# show   ip   route
Total route items is 3, the matched route items is 3
Codes: C - connected, S - static, R - RIP derived, O - OSPF derived
       A - OSPF ASE, B - BGP derived, D - DVMRP derived

Destination	Mask	Nexthop	Interface	Preference
C 192.168.10.0	255.255.255.0	0.0.0.0	Vlan10	0
C 192.168.20.0	255.255.255.0	0.0.0.0	Vlan20	0
C 192.168.30.0	255.255.255.0	0.0.0.0	Vlan30	0

第七步：对 PC1 和 PC2，进行以下验证实验。
1. "本地连接 2"都不配网关， PC1 和 PC2 彼此 ping 不通。
2. "本地连接 2"都配置网关， PC1 和 PC2 彼此能 ping 通。

实验十五　多层交换机静态路由实验

一、实验目的

1. 理解三层交换机进行路由的原理和具体实现拓扑。
2. 理解三层交换机静态路由的配置方法。

二、相关知识

多层交换机具备第二层交换(基于 MAC 地址)和第三层交换(基于 IP 地址)的能力,能够同时实现静态路由并保持高交换速度。静态路由可以通过在多层交换机上配置静态路由表来实现，这通常包括设置静态路由的目的地址、下一跳地址等参数。

多层交换机静态路由实验应用场景是：当两台三层交换机级联时，为了保证每台交换机上所连接的网段可以和另一台交换机上连接的网段互相通信，最简单的方法就是配置静态路由。

三、实验设备

1. DCRS-7604(或 6804)交换机 1 台。
2. DCRS-5526S 交换机 1 台。
3. PC 机 2~4 台。
4. Console 线 1~2 根。
5. 直通网线 2~4 根。

四、实验拓扑

该实验拓扑结构如图 1-40 所示。

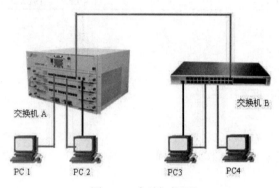

图 1-40 实验拓扑图

五、实验要求

1. 在交换机 A 和交换机 B 上分别划分基于端口的 VLAN。
2. 交换机 A 和交换机 B 通过 24 口进行级联。
3. 配置交换机 A 和交换机 B 各 VLAN 虚拟接口的 IP 地址,如表 1-30、表 1-31 和表 1-32 所示。

表 1-30 交换机 A 和交换机 B 的 VLAN 划分

交换机	VLAN	端口成员
交换机 A	10	1~8
	20	9~16
	100	24
交换机 B	30	1~8
	40	9~16
	101	24

表 1-31　交换机 A 和交换机 B 的 IP 配置

设备	IP 地址	gateway	Mask
PC1	192.168.10.101	192.168.10.1	255.255.255.0
PC2	192.168.20.101	192.168.20.1	255.255.255.0
PC3	192.168.30.101	192.168.30.1	255.255.255.0
PC4	192.168.40.101	192.168.40.1	255.255.255.0

表 1-32　PC 的配置

VLAN 10	VLAN 20	VLAN 30	VLAN 40	VLAN 100	VLAN 101
192.168.10.1	192.168.20.1	192.168.30.1	192.168.40.1	192.168.100.1	192.168.100.2

验证结果注意下列问题。

(1) 没有静态路由之前：

PC1、PC2 与 PC3、PC4 可以互通。

PC1、PC2 与 PC3、PC4 不通。

(2) 配置静态路由之后：

4 台 PC 之间都可以互通。

若实验结果和理论相符，则本实验完成。

六、实验步骤

神州数码设备的配置步骤如下。

第一步：交换机全部恢复出厂设置，配置交换机的 VLAN 信息。

交换机 A：

```
DCRS-7604#conf
DCRS-7604(Config)#vlan 10
DCRS-7604(Config-Vlan10)#switchport interface ethernet 1/1-8
Set the port Ethernet1/1 access vlan 10 successfully
Set the port Ethernet1/2 access vlan 10 successfully
Set the port Ethernet1/3 access vlan 10 successfully
Set the port Ethernet1/4 access vlan 10 successfully
Set the port Ethernet1/5 access vlan 10 successfully
Set the port Ethernet1/6 access vlan 10 successfully
Set the port Ethernet1/7 access vlan 10 successfully
Set the port Ethernet1/8 access vlan 10 successfully
DCRS-7604(Config-Vlan10)#exit
DCRS-7604(Config)#vlan 20
DCRS-7604(Config-Vlan20)#switchport interface ethernet 1/9-16
Set the port Ethernet1/9 access vlan 20 successfully
Set the port Ethernet1/10 access vlan 20 successfully
```

Set the port Ethernet1/11 access vlan 20 successfully
Set the port Ethernet1/12 access vlan 20 successfully
Set the port Ethernet1/13 access vlan 20 successfully
Set the port Ethernet1/14 access vlan 20 successfully
Set the port Ethernet1/15 access vlan 20 successfully
Set the port Ethernet1/16 access vlan 20 successfully
DCRS-7604(Config-Vlan20)#exit
DCRS-7604(Config)#vlan 100
DCRS-7604(Config-Vlan100)#switchport interface ethernet 1/24
Set the port Ethernet1/24 access vlan 100 successfully
DCRS-7604(Config-Vlan100)#exit
DCRS-7604(Config)#

验证配置：

```
DCRS-7604#show vlan
VLAN  Name       Type     Media   Ports
------------------------------------------------------------
1     default    Static   ENET    Ethernet1/17   Ethernet1/18
                                  Ethernet1/19   Ethernet1/20
                                  Ethernet1/21   Ethernet1/22
                                  Ethernet1/23   Ethernet1/25
                                  Ethernet1/26   Ethernet1/27
                                  Ethernet1/28
10    VLAN0010   Static   ENET    Ethernet1/1    Ethernet1/2
                                  Ethernet1/3    Ethernet1/4
                                  Ethernet1/5    Ethernet1/6
                                  Ethernet1/7    Ethernet1/8
20    VLAN0020   Static   ENET    Ethernet1/9    Ethernet1/10
                                  Ethernet1/11   Ethernet1/12
                                  Ethernet1/13   Ethernet1/14
                                  Ethernet1/15   Ethernet1/16
100   VLAN0100   Static   ENET    Ethernet1/24
DCRS-7604#
```

交换机 B：

DCRS-5526S(Config)#vlan 30
DCRS-5526S(Config-Vlan30)#switchport interface ethernet 0/0/1-8
Set the port Ethernet0/0/1 access vlan 30 successfully
Set the port Ethernet0/0/2 access vlan 30 successfully
Set the port Ethernet0/0/3 access vlan 30 successfully
Set the port Ethernet0/0/4 access vlan 30 successfully
Set the port Ethernet0/0/5 access vlan 30 successfully

```
Set the port Ethernet0/0/6 access vlan 30 successfully
Set the port Ethernet0/0/7 access vlan 30 successfully
Set the port Ethernet0/0/8 access vlan 30 successfully
DCRS-5526S(Config-Vlan30)#exit
DCRS-5526S(Config)#vlan 40
DCRS-5526S(Config-Vlan40)#switchport interface ethernet 0/0/9-16
Set the port Ethernet0/0/9 access vlan 40 successfully
Set the port Ethernet0/0/10 access vlan 40 successfully
Set the port Ethernet0/0/11 access vlan 40 successfully
Set the port Ethernet0/0/12 access vlan 40 successfully
Set the port Ethernet0/0/13 access vlan 40 successfully
Set the port Ethernet0/0/14 access vlan 40 successfully
Set the port Ethernet0/0/15 access vlan 40 successfully
Set the port Ethernet0/0/16 access vlan 40 successfully
DCRS-5526S(Config-Vlan40)#exit
DCRS-5526S(Config)#vlan 101
DCRS-5526S(Config-Vlan101)#switchport interface ethernet 0/0/24
Set the port Ethernet0/0/24 access vlan 101 successfully
DCRS-5526S(Config-Vlan101)#exit
DCRS-5526S(Config)#
```

验证配置：

```
DCRS-5526S#show vlan
VLAN  Name       Type     Media   Ports
------------------------------  --------------------------------
1     default    Static   ENET    Ethernet0/0/17   Ethernet0/0/18
                                  Ethernet0/0/19   Ethernet0/0/20
                                  Ethernet0/0/21   Ethernet0/0/22
                                  Ethernet0/0/23
30    VLAN0030   Static   ENET    Ethernet0/0/1    Ethernet0/0/2
                                  Ethernet0/0/3    Ethernet0/0/4
                                  Ethernet0/0/5    Ethernet0/0/6
                                  Ethernet0/0/7    Ethernet0/0/8
40    VLAN0040   Static   ENET    Ethernet0/0/9    Ethernet0/0/10
                                  Ethernet0/0/11   Ethernet0/0/12
                                  Ethernet0/0/13   Ethernet0/0/14
                                  Ethernet0/0/15   Ethernet0/0/16
101   VLAN0101   Static   ENET    Ethernet0/0/24
DCRS-5526S#
```

第二步：配置交换机各 VLAN 虚接口的 IP 地址。

交换机 A：

DCRS-7604(Config)#int vlan 10
DCRS-7604(Config-If-Vlan10)#ip address 192.168.10.1 255.255.255.0
DCRS-7604(Config-If-Vlan10)#no shut
DCRS-7604(Config-If-Vlan10)#exit
DCRS-7604(Config)#int vlan 20
DCRS-7604(Config-If-Vlan20)#ip address 192.168.20.1 255.255.255.0
DCRS-7604(Config-If-Vlan20)#no shut
DCRS-7604(Config-If-Vlan20)#exit
DCRS-7604(Config)#int vlan 100
DCRS-7604(Config-If-Vlan100)#ip address 192.168.100.1 255.255.255.0
DCRS-7604(Config-If-Vlan100)#no shut
DCRS-7604(Config-If-Vlan100)#
DCRS-7604(Config-If-Vlan100)#exit
DCRS-7604(Config)#

交换机 B：

DCRS-5526S(Config)#int vlan 30
DCRS-5526S(Config-If-Vlan30)#ip address 192.168.30.1 255.255.255.0
DCRS-5526S(Config-If-Vlan30)#no shut
DCRS-5526S(Config-If-Vlan30)#exit
DCRS-5526S(Config)#interface vlan 40
DCRS-5526S(Config-If-Vlan40)#ip address 192.168.40.1 255.255.255.0
DCRS-5526S(Config-If-Vlan40)#exit
DCRS-5526S(Config)#int vlan 101
DCRS-5526S(Config-If-Vlan101)#ip address 192.168.100.2 255.255.255.0
DCRS-5526S(Config-If-Vlan101)#exit
DCRS-5526S(Config)#

第三步：配置 PC 的 IP 地址(注意配置网关)，如表 1-33 所示。

表 1-33 PC 的 IP 地址配置

设备	IP 地址	gateway	Mask
PC1	192.168.10.101	192.168.10.1	255.255.255.0
PC2	192.168.20.101	192.168.20.1	255.255.255.0
PC3	192.168.30.101	192.168.30.1	255.255.255.0
PC4	192.168.40.101	192.168.40.1	255.255.255.0

第四步：验证 PC 之间的连通性，如表 1-34 所示。

表 1-34 验证 PC 之间的连通性

PC	端口	PC	端口	结果	原因
PC1	A: 1/1	PC2	A: 1/9	通	
PC1	A: 1/1	VLAN 100	A: 1/24	通	
PC1	A: 1/1	VLAN 101	B: 0/0/24	不通	
PC1	A: 1/1	PC3	B: 0/0/1	不通	

查看路由表，进一步分析上一步的现象原因。

交换机 A：

```
DCRS-7604#show ip route
Total route items is 3, the matched route items is 3
Codes: C - connected, S - static, R - RIP derived, O - OSPF derived
       A - OSPF ASE, B - BGP derived, D - DVMRP derived
Destination        Mask              Nexthop        Interface      Preference
C   192.168.10.0   255.255.255.0     0.0.0.0        Vlan10         0
C   192.168.20.0   255.255.255.0     0.0.0.0        Vlan20         0
C   192.168.100.0  255.255.255.0     0.0.0.0        Vlan100        0
DCRS-7604#
```

交换机 B：

```
DCRS-5526S#show ip route
Total route items is 3, the matched route items is 3
Codes: C - connected, S - static, R - RIP derived, O - OSPF derived
       A - OSPF ASE, B - BGP derived, D - DVMRP derived
Destination        Mask              Nexthop        Interface      Preference
C   192.168.30.0   255.255.255.0     0.0.0.0        Vlan30         0
C   192.168.40.0   255.255.255.0     0.0.0.0        Vlan40         0
C   192.168.100.0  255.255.255.0     0.0.0.0        Vlan101        0
DCRS-5526S#
```

第五步：配置静态路由。

交换机 A：

```
DCRS-7604(Config)#ip route 192.168.30.0 255.255.255.0 192.168.100.2
DCRS-7604(Config)#ip route 192.168.40.0 255.255.255.0 192.168.100.2
```

验证配置：

```
DCRS-7604#show ip route
Total route items is 5, the matched route items is 5
Codes: C - connected, S - static, R - RIP derived, O - OSPF derived
       A - OSPF ASE, B - BGP derived, D - DVMRP derived
```

Destination	Mask	Nexthop	Interface	Preference
C 192.168.10.0	255.255.255.0	0.0.0.0	Vlan10	0
C 192.168.20.0	255.255.255.0	0.0.0.0	Vlan20	0
S 192.168.30.0	255.255.255.0	192.168.100.2	Vlan100	1
S 192.168.40.0	255.255.255.0	192.168.100.2	Vlan100	1
C 192.168.100.0	255.255.255.0	0.0.0.0	Vlan100	0

DCRS-7604#

交换机 B：

DCRS-5526S(Config)#ip route 192.168.10.0 255.255.255.0 192.168.100.1
DCRS-5526S(Config)#ip route 192.168.20.0 255.255.255.0 192.168.100.1

验证配置：

DCRS-5526S#show ip route
Total route items is 5, the matched route items is 5
Codes: C - connected, S - static, R - RIP derived, O - OSPF derived
       A - OSPF ASE, B - BGP derived, D - DVMRP derived

Destination	Mask	Nexthop	Interface	Preference
S 192.168.10.0	255.255.255.0	192.168.100.1	Vlan101	1
S 192.168.20.0	255.255.255.0	192.168.100.1	Vlan101	1
C 192.168.30.0	255.255.255.0	0.0.0.0	Vlan30	0
C 192.168.40.0	255.255.255.0	0.0.0.0	Vlan40	0
C 192.168.100.0	255.255.255.0	0.0.0.0	Vlan101	0

DCRS-5526S#

思科设备的配置命令如下。

三层交换机 RS3560-A：

Switch>enable
Switch#configure terminal
Switch(config)#hostname RS3560-A
RS3560-A(config)#
RS3560-A(config)#vlan 10
RS3560-A(config-vlan)#ex
RS3560-A(config)#vlan 20
RS3560-A(config-vlan)#ex
RS3560-A(config)#vlan 100
RS3560-A(config-vlan)#ex
RS3560-A(config)#interface range fastEthernet 0/1-8
RS3560-A(config-if-range)#switchport access vlan 10
RS3560-A(config-if-range)#ex
RS3560-A(config)#interface range fastEthernet 0/9-16

RS3560-A(config-if-range)#switchport access vlan 20
RS3560-A(config-if-range)#ex
RS3560-A(config)#interface fastEthernet 0/24
RS3560-A(config-if)#switchport access vlan 100
RS3560-A(config-if)#ex
RS3560-A(config)#
RS3560-A(config)#interface vlan 10
RS3560-A(config-if)#
%LINK-5-CHANGED: Interface Vlan10, changed state to up
RS3560-A(config-if)#ip address 192.168.10.1 255.255.255.0
RS3560-A(config-if)#no shutdown
RS3560-A(config-if)#interface vlan 20
RS3560-A(config-if)#
%LINK-5-CHANGED: Interface Vlan20, changed state to up
RS3560-A(config-if)#ip address 192.168.20.1 255.255.255.0
RS3560-A(config-if)#no shutdown
RS3560-A(config-if)#interface vlan 100
RS3560-A(config-if)#
%LINK-5-CHANGED: Interface Vlan100, changed state to up
RS3560-A(config-if)#ip address 192.168.100.1 255.255.255.0
RS3560-A(config-if)#no shutdown
RS3560-A(config-if)#ex
RS3560-A(config)#ip routing
RS3560-A(config)#
RS3560-A#sh ip route
Codes: C - connected, S - static, I - IGRP, R - RIP, M - mobile, B - BGP
       D - EIGRP, EX - EIGRP external, O - OSPF, IA - OSPF inter area
       N1 - OSPF NSSA external type 1, N2 - OSPF NSSA external type 2
       E1 - OSPF external type 1, E2 - OSPF external type 2, E - EGP
       i - IS-IS, L1 - IS-IS level-1, L2 - IS-IS level-2, ia - IS-IS inter area
       * - candidate default, U - per-user static route, o - ODR
       P - periodic downloaded static route

Gateway of last resort is not set

C   192.168.10.0/24 is directly connected, Vlan10
C   192.168.20.0/24 is directly connected, Vlan20
C   192.168.100.0/24 is directly connected, Vlan100

三层交换机 RS3560-B：

Switch>enable
Switch#configure terminal
Switch(config)#hostname RS3560-B
RS3560-B(config)#vl

```
RS3560-B(config)#vlan 30
RS3560-B(config-vlan)#ex
RS3560-B(config)#vlan 40
RS3560-B(config-vlan)#ex
RS3560-B(config)#vlan 101
RS3560-B(config-vlan)#ex
RS3560-B(config)#interface range fastEthernet 0/1-8
RS3560-B(config-if-range)#switchport access vlan 30
RS3560-B(config-if-range)#interface range fastEthernet 0/9-16
RS3560-B(config-if-range)#switchport access vlan 40
RS3560-B(config-if-range)#ex
RS3560-B(config)#interface fastEthernet 0/24
RS3560-B(config-if)#switchport access vlan 101
RS3560-B(config)#interface vlan 30
RS3560-B(config-if)#
%LINK-5-CHANGED: Interface Vlan30, changed state to up
RS3560-B(config-if)#ip address 192.168.30.1 255.255.255.0
RS3560-B(config-if)#no shutdown
RS3560-B(config-if)#interface vlan 40
%LINK-5-CHANGED: Interface Vlan40, changed state to up
RS3560-B(config-if)#ip address 192.168.40.1 255.255.255.0
RS3560-B(config-if)#no shutdown
RS3560-B(config-if)#interface vlan 101
%LINK-5-CHANGED: Interface Vlan101, changed state to up
RS3560-B(config-if)#ip address 192.168.100.2 255.255.255.0
RS3560-B(config-if)#no shutdown
RS3560-B(config-if)#ex
RS3560-B(config)#ip routing
RS3560-B(config)#
RS3560-B#sh ip route
Codes: C - connected, S - static, I - IGRP, R - RIP, M - mobile, B - BGP
       D - EIGRP, EX - EIGRP external, O - OSPF, IA - OSPF inter area
       N1 - OSPF NSSA external type 1, N2 - OSPF NSSA external type 2
       E1 - OSPF external type 1, E2 - OSPF external type 2, E - EGP
       i - IS-IS, L1 - IS-IS level-1, L2 - IS-IS level-2, ia - IS-IS inter area
       * - candidate default, U - per-user static route, o - ODR
       P - periodic downloaded static route
Gateway of last resort is not set
C    192.168.30.0/24 is directly connected, Vlan30
C    192.168.40.0/24 is directly connected, Vlan40
C    192.168.100.0/24 is directly connected, Vlan101
```

配置静态路由。
三层交换机 RS3560-A：

RS3560-A(config)#ip route 192.168.30.0 255.255.255.0 192.168.100.2
RS3560-A(config)#ip route 192.168.40.0 255.255.255.0 192.168.100.2
RS3560-A(config)#ex
RS3560-A#
%SYS-5-CONFIG_I: Configured from console by console
RS3560-A#sh ip route
Codes: C - connected, S - static, I - IGRP, R - RIP, M - mobile, B - BGP
       D - EIGRP, EX - EIGRP external, O - OSPF, IA - OSPF inter area
       N1 - OSPF NSSA external type 1, N2 - OSPF NSSA external type 2
       E1 - OSPF external type 1, E2 - OSPF external type 2, E - EGP
       i - IS-IS, L1 - IS-IS level-1, L2 - IS-IS level-2, ia - IS-IS inter area
       * - candidate default, U - per-user static route, o - ODR
       P - periodic downloaded static route

Gateway of last resort is not set

C   192.168.10.0/24 is directly connected, Vlan10
C   192.168.20.0/24 is directly connected, Vlan20
S   192.168.30.0/24 [1/0] via 192.168.100.2
S   192.168.40.0/24 [1/0] via 192.168.100.2
C   192.168.100.0/24 is directly connected, Vlan100

三层交换机 RS3560-B：

RS3560-B(config)#ip route 192.168.10.0 255.255.255.0 192.168.100.1
RS3560-B(config)#ip route 192.168.20.0 255.255.255.0 192.168.100.1
RS3560-B(config)#^Z
RS3560-B#
%SYS-5-CONFIG_I: Configured from console by console
RS3560-B#
RS3560-B#sh ip route
Codes: C - connected, S - static, I - IGRP, R - RIP, M - mobile, B - BGP
       D - EIGRP, EX - EIGRP external, O - OSPF, IA - OSPF inter area
       N1 - OSPF NSSA external type 1, N2 - OSPF NSSA external type 2
       E1 - OSPF external type 1, E2 - OSPF external type 2, E - EGP
       i - IS-IS, L1 - IS-IS level-1, L2 - IS-IS level-2, ia - IS-IS inter area
       * - candidate default, U - per-user static route, o - ODR
       P - periodic downloaded static route

Gateway of last resort is not set

S   192.168.10.0/24 [1/0] via 192.168.100.1
S   192.168.20.0/24 [1/0] via 192.168.100.1
C   192.168.30.0/24 is directly connected, Vlan30

| C | 192.168.40.0/24 is directly connected, Vlan40 |
| C | 192.168.100.0/24 is directly connected, Vlan101 |

配置序列：

RS3560-A#sh run
Building configuration...
Current configuration : 1874 bytes
!
version 12.2
no service timestamps log datetime msec
no service timestamps debug datetime msec
no service password-encryption
!
hostname RS3560-A
!
ip routing
!
spanning-tree mode pvst
!
interface FastEthernet0/1
 switchport access vlan 10
!
interface FastEthernet0/2
 switchport access vlan 10
!
interface FastEthernet0/3
 switchport access vlan 10
!
interface FastEthernet0/4
 switchport access vlan 10
!
interface FastEthernet0/5
 switchport access vlan 10
!
interface FastEthernet0/6
 switchport access vlan 10
!
interface FastEthernet0/7
 switchport access vlan 10
!
interface FastEthernet0/8
 switchport access vlan 10

!
interface FastEthernet0/9
  switchport access vlan 20
!
interface FastEthernet0/10
  switchport access vlan 20
!
interface FastEthernet0/11
  switchport access vlan 20
!
interface FastEthernet0/12
  switchport access vlan 20
!
interface FastEthernet0/13
  switchport access vlan 20
!
interface FastEthernet0/14
  switchport access vlan 20
!
interface FastEthernet0/15
  switchport access vlan 20
!
interface FastEthernet0/16
  switchport access vlan 20
!
interface FastEthernet0/17
!
interface FastEthernet0/18
!
interface FastEthernet0/19
!
interface FastEthernet0/20
!
interface FastEthernet0/21
!
interface FastEthernet0/22
!
interface FastEthernet0/23
!
interface FastEthernet0/24
  switchport access vlan 100
!

```
interface GigabitEthernet0/1
!
interface GigabitEthernet0/2
!
interface Vlan1
 no ip address
 shutdown
!
interface Vlan10
 ip address 192.168.10.1 255.255.255.0
!
interface Vlan20
 ip address 192.168.20.1 255.255.255.0
!
interface Vlan100
 ip address 192.168.100.1 255.255.255.0
!
ip classless
ip route 192.168.30.0 255.255.255.0 192.168.100.2
ip route 192.168.40.0 255.255.255.0 192.168.100.2
!
ip flow-export version 9
!
line con 0
!
line aux 0
!
line vty 0 4
 login
!
end

RS3560-A#

RS3560-B#sh run
Building configuration...
Current configuration : 1874 bytes
!
version 12.2
no service timestamps log datetime msec
no service timestamps debug datetime msec
no service password-encryption
```

```
!
hostname RS3560-B
!
ip routing
!
spanning-tree mode pvst
!
interface FastEthernet0/1
  switchport access vlan 30
!
interface FastEthernet0/2
  switchport access vlan 30
!
interface FastEthernet0/3
  switchport access vlan 30
!
interface FastEthernet0/4
  switchport access vlan 30
!
interface FastEthernet0/5
  switchport access vlan 30
!
interface FastEthernet0/6
  switchport access vlan 30
!
interface FastEthernet0/7
  switchport access vlan 30
!
interface FastEthernet0/8
  switchport access vlan 30
!
interface FastEthernet0/9
  switchport access vlan 40
!
interface FastEthernet0/10
  switchport access vlan 40
!
interface FastEthernet0/11
  switchport access vlan 40
!
interface FastEthernet0/12
  switchport access vlan 40
```

```
!
interface FastEthernet0/13
  switchport access vlan 40
!
interface FastEthernet0/14
  switchport access vlan 40
!
interface FastEthernet0/15
  switchport access vlan 40
!
interface FastEthernet0/16
  switchport access vlan 40
!
interface FastEthernet0/17
!
interface FastEthernet0/18
!
interface FastEthernet0/19
!
interface FastEthernet0/20
!
interface FastEthernet0/21
!
interface FastEthernet0/22
!
interface FastEthernet0/23
!
interface FastEthernet0/24
  switchport access vlan 101
!
interface GigabitEthernet0/1
!
interface GigabitEthernet0/2
!
interface Vlan1
  no ip address
  shutdown
!
interface Vlan30
  ip address 192.168.30.1 255.255.255.0
!
interface Vlan40
```

```
  ip address 192.168.40.1 255.255.255.0
!
interface Vlan101
  ip address 192.168.100.2 255.255.255.0
!
ip classless
ip route 192.168.10.0 255.255.255.0 192.168.100.1
ip route 192.168.20.0 255.255.255.0 192.168.100.1
!
ip flow-export version 9
!
line con 0
!
line aux 0
!
line vty 0 4
  login
!
!
end
RS3560-B#
```

第六步：验证 PC 之间的连通性，如表 1-35 所示。

表 1-35　验证 PC 之间的连通性

PC	端口	PC	端口	结果	原因
PC1	A：1/1	PC2	A：1/9	通	
PC1	A：1/1	VLAN 100	A：1/24	通	
PC1	A：1/1	VLAN 101	B：0/0/24	通	
PC1	A：1/1	PC3	B：0/0/1	通	

七、注意事项和排错

1. PC 一定要配置正确的网关，否则不能正常通信。

2. 两台交换机级联的端口可以在同一 VLAN，也可以在不同 VLAN。

八、思考题

1. 如果将交换机 B 上的 VLAN 30 改为 VLAN 10，两台交换机上的 VLAN 10 是否为同一个 VLAN？

2. 本次实验第四步中，PC1 ping VLAN 101 以及 PC1 ping PC3 都不通，原因是什么？

实训项目五 多层交换机静态路由配置

一、实训目的

1. 学会在多层交换机上配置静态路由，实现网络中不同 VLAN 或不同子网之间的通信。

2. 验证静态路由配置的有效性，诊断并解决静态路由配置问题。

二、实训设备

(1) 二层 S 交换机 2 台。

(2) 三层 RS 交换机 2 台。

(3) PC 机 4 台。

(4) 直通网线 7 根。

三、实训拓扑

该实验拓扑结构如图 1-41 所示。

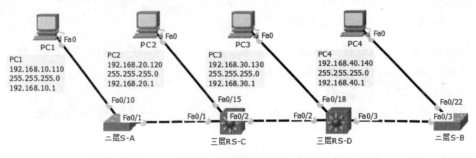

图 1-41 实验拓扑图

四、实训要求

1. 参照图 1-41 所示进行连线。

2. 实训结果：PC1---ping---PC2----PC3----PC4   全通。

五、实训步骤

第一步：对交换机 A、B、C 和 D 进行配置。

交换机 A：划分 vl 10(9~12)、vl 20(13~16)，Trunk (1)。

交换机 B：划分 vl 30(17~20)、vl 40(21~24)，Trunk (3)。

交换机 C：划分 vl 10(9~12)、vl 20(13~16)、vl 100(2)，Trunk (1)。

交换机 D：划分 vl 30(17~20)、vl 40(21~24)、vl 101(2)，Trunk (3)。

第二步：设置三层交换机 C 启用三层路由功能。
设置 vl 10、vl 20、vl 100 的接口地址：

int vl 10   192.168.10.1/24；
int vl 20   192.168.20.1/24；
int vl 100   192.168.100.1/30。

设置三层交换机 D 启用三层路由功能。
设置 vl 30、vl 40、vl 101 的接口地址：

int vl 30   192.168.30.1/24；
int vl 40   192.168.40.1/24；
int vl 101   192.168.100.2/30。

第三步：配置交换机 C 的静态路由。

C(config)#ip route 192.168.30.0 255.255.255.0 192.168.100.2
C(config)#ip route 192.168.40.0 255.255.255.0 192.168.100.2

配置交换机 D 的静态路由。

D(config)#ip route 192.168.10.0 255.255.255.0 192.168.100.1
D(config)#ip route 192.168.20.0 255.255.255.0 192.168.100.1

第四步：根据连线位置正确配置 PC1、PC2、PC3、PC4 的地址(注意配置网关地址,并确保其与 VLAN 的对应关系正确匹配)。

实验结果：PC1---ping---PC2----PC3----PC4   全通。
查看交换机状态：sh vlan   sh run   sh ip route。

实验十六 三层交换机 RIP 动态路由

一、实验目的

1. 掌握三层交换机之间通过 RIP 协议实现网段互通的配置方法。
2. 理解动态实现方式与静态方式的不同。

二、相关知识

RIP(Routing Information Protocol)是一种基于距离向量的动态路由协议，用于在自治系统内部交换路由信息。当两台三层交换机级联时，为了保证每台交换机上连接的网段可以和另一台交换机上连接的网段互通，使用 RIP 协议可以动态学习路由，从而保持高效的交换速度。

应用场景如下。

(1) 校园网络：在校园网络中，不同部门可能需要访问不同的资源。RIP 动态路由可以实现资源的隔离和访问控制。

(2) 企业网络管理：在企业网络中，通过 RIP 动态路由可以实现网络的自动管理和控制。

(3) 数据中心网络：在数据中心中，服务器和存储设备可能分布在不同的 VLAN 中。RIP 动态路由可以实现这些设备之间的通信。

三、实验设备

1. DCRS-7604(或 6804)交换机 1 台。
2. DCRS-5526S 交换机 1 台。
3. PC 机 2~4 台。
4. Console 线 1~2 根。
5. 直通网线 2~4 根。

四、实验拓扑

该实验拓扑结构如图 1-42 所示。

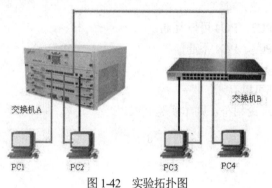

图 1-42 实验拓扑图

五、实验要求

1. 参照表 1-36 所示在交换机 A 和交换机 B 上分别划分基于端口的 VLAN。

表 1-36 交换机 A 和交换机 B 的 VLAN 划分

交换机	VLAN	端口成员
交换机 A	10	1~8
	20	9~16
	100	24
交换机 B	30	1~8
	40	9~16
	101	24

2. 交换机 A 和交换机 B 通过 24 口进行级联。

3. 参照如表 1-37 所示配置交换机 A 和交换机 B 各 VLAN 虚拟接口的 IP 地址。

表 1-37　交换机 A 和交换机 B 的 IP 配置

VLAN 10	VLAN 20	VLAN 30	VLAN 40	VLAN 100	VLAN 101
192.168.10.1	192.168.20.1	192.168.30.1	192.168.40.1	192.168.100.1	192.168.100.2

4. PC 的配置如表 1-38 所示。

表 1-38　PC 的配置

设备	IP 地址	gateway	Mask
PC1	192.168.10.101	192.168.10.1	255.255.255.0
PC2	192.168.20.101	192.168.20.1	255.255.255.0
PC3	192.168.30.101	192.168.30.1	255.255.255.0
PC4	192.168.40.101	192.168.40.1	255.255.255.0

验证结果注意下列问题。

(1) 没有 RIP 路由协议之前：

　　PC1、PC2 与 PC3、PC4 可以互通。

　　PC1、PC2 与 PC3、PC4 不通。

(2) 配置 RIP 路由协议之后：

　　4 台 PC 之间都可以互通。

　　若实验结果和理论相符，则本实验完成。

六、实验步骤

神州数码设备的配置步骤如下。

第一步：交换机全部恢复出厂设置，配置交换机的 VLAN 信息。

交换机 A：

```
DCRS-7604#conf
DCRS-7604(Config)#vlan 10
DCRS-7604(Config-Vlan10)#switchport interface ethernet 1/1-8
Set the port Ethernet1/1 access vlan 10 successfully
Set the port Ethernet1/2 access vlan 10 successfully
Set the port Ethernet1/3 access vlan 10 successfully
Set the port Ethernet1/4 access vlan 10 successfully
Set the port Ethernet1/5 access vlan 10 successfully
Set the port Ethernet1/6 access vlan 10 successfully
Set the port Ethernet1/7 access vlan 10 successfully
Set the port Ethernet1/8 access vlan 10 successfully
DCRS-7604(Config-Vlan10)#exit
```

```
DCRS-7604(Config)#vlan 20
DCRS-7604(Config-Vlan20)#switchport interface ethernet 1/9-16
Set the port Ethernet1/9 access vlan 20 successfully
Set the port Ethernet1/10 access vlan 20 successfully
Set the port Ethernet1/11 access vlan 20 successfully
Set the port Ethernet1/12 access vlan 20 successfully
Set the port Ethernet1/13 access vlan 20 successfully
Set the port Ethernet1/14 access vlan 20 successfully
Set the port Ethernet1/15 access vlan 20 successfully
Set the port Ethernet1/16 access vlan 20 successfully
DCRS-7604(Config-Vlan20)#exit
DCRS-7604(Config)#vlan 100
DCRS-7604(Config-Vlan100)#switchport interface ethernet 1/24
Set the port Ethernet1/24 access vlan 100 successfully
DCRS-7604(Config-Vlan100)#exit
DCRS-7604(Config)#
```

验证配置：

VLAN	Name	Type	Media	Ports	
DCRS-7604#show vlan					
1	default	Static	ENET	Ethernet1/17	Ethernet1/18
				Ethernet1/19	Ethernet1/20
				Ethernet1/21	Ethernet1/22
				Ethernet1/23	Ethernet1/25
				Ethernet1/26	Ethernet1/27
				Ethernet1/28	
10	VLAN0010	Static	ENET	Ethernet1/1	Ethernet1/2
				Ethernet1/3	Ethernet1/4
				Ethernet1/5	Ethernet1/6
				Ethernet1/7	Ethernet1/8
20	VLAN0020	Static	ENET	Ethernet1/9	Ethernet1/10
				Ethernet1/11	Ethernet1/12
				Ethernet1/13	Ethernet1/14
				Ethernet1/15	Ethernet1/16
100	VLAN0100	Static	ENET	Ethernet1/24	
DCRS-7604#					

交换机 B：

```
DCRS-5526S(Config)#vlan 30
DCRS-5526S(Config-Vlan30)#switchport interface ethernet 0/0/1-8
Set the port Ethernet0/0/1 access vlan 30 successfully
Set the port Ethernet0/0/2 access vlan 30 successfully
Set the port Ethernet0/0/3 access vlan 30 successfully
Set the port Ethernet0/0/4 access vlan 30 successfully
```

Set the port Ethernet0/0/5 access vlan 30 successfully
Set the port Ethernet0/0/6 access vlan 30 successfully
Set the port Ethernet0/0/7 access vlan 30 successfully
Set the port Ethernet0/0/8 access vlan 30 successfully
DCRS-5526S(Config-Vlan30)#exit
DCRS-5526S(Config)#vlan 40
DCRS-5526S(Config-Vlan40)#switchport interface ethernet 0/0/9-16
Set the port Ethernet0/0/9 access vlan 40 successfully
Set the port Ethernet0/0/10 access vlan 40 successfully
Set the port Ethernet0/0/11 access vlan 40 successfully
Set the port Ethernet0/0/12 access vlan 40 successfully
Set the port Ethernet0/0/13 access vlan 40 successfully
Set the port Ethernet0/0/14 access vlan 40 successfully
Set the port Ethernet0/0/15 access vlan 40 successfully
Set the port Ethernet0/0/16 access vlan 40 successfully
DCRS-5526S(Config-Vlan40)#exit
DCRS-5526S(Config)#vlan 101
DCRS-5526S(Config-Vlan101)#switchport interface ethernet 0/0/24
Set the port Ethernet0/0/24 access vlan 101 successfully
DCRS-5526S(Config-Vlan101)#exit
DCRS-5526S(Config)#

验证配置：

DCRS-5526S#show vlan

VLAN	Name	Type	Media	Ports	
1	default	Static	ENET	Ethernet0/0/17	Ethernet0/0/18
				Ethernet0/0/19	Ethernet0/0/20
				Ethernet0/0/21	Ethernet0/0/22
				Ethernet0/0/23	
30	VLAN0030	Static	ENET	Ethernet0/0/1	Ethernet0/0/2
				Ethernet0/0/3	Ethernet0/0/4
				Ethernet0/0/5	Ethernet0/0/6
				Ethernet0/0/7	Ethernet0/0/8
40	VLAN0040	Static	ENET	Ethernet0/0/9	Ethernet0/0/10
				Ethernet0/0/11	Ethernet0/0/12
				Ethernet0/0/13	Ethernet0/0/14
				Ethernet0/0/15	Ethernet0/0/16
101	VLAN0101	Static	ENET	Ethernet0/0/24	

DCRS-5526S#

第二步：配置交换机各 VLAN 虚接口的 IP 地址。

交换机 A：

DCRS-7604(Config)#int vlan 10
DCRS-7604(Config-If-Vlan10)#ip address 192.168.10.1 255.255.255.0

```
DCRS-7604(Config-If-Vlan10)#no shut
DCRS-7604(Config-If-Vlan10)#exit
DCRS-7604(Config)#int vlan 20
DCRS-7604(Config-If-Vlan20)#ip address 192.168.20.1 255.255.255.0
DCRS-7604(Config-If-Vlan20)#no shut
DCRS-7604(Config-If-Vlan20)#exit
DCRS-7604(Config)#int vlan 100
DCRS-7604(Config-If-Vlan100)#ip address 192.168.100.1 255.255.255.0
DCRS-7604(Config-If-Vlan100)#no shut
DCRS-7604(Config-If-Vlan100)#
DCRS-7604(Config-If-Vlan100)#exit
DCRS-7604(Config)#
```

交换机 B：

```
DCRS-5526S(Config)#int vlan 30
DCRS-5526S(Config-If-Vlan30)#ip address 192.168.30.1 255.255.255.0
DCRS-5526S(Config-If-Vlan30)#no shut
DCRS-5526S(Config-If-Vlan30)#exit
DCRS-5526S(Config)#interface vlan 40
DCRS-5526S(Config-If-Vlan40)#ip address 192.168.40.1 255.255.255.0
DCRS-5526S(Config-If-Vlan40)#exit
DCRS-5526S(Config)#int vlan 101
DCRS-5526S(Config-If-Vlan101)#ip address 192.168.100.2 255.255.255.0
DCRS-5526S(Config-If-Vlan101)#exit
DCRS-5526S(Config)#
```

第三步：参照如表 1-39 所示配置 PC 的 IP 地址(注意配置网关)。

表 1-39　PC 的 IP 地址配置

设备	IP 地址	gateway	Mask
PC1	192.168.10.101	192.168.10.1	255.255.255.0
PC2	192.168.20.101	192.168.20.1	255.255.255.0
PC3	192.168.30.101	192.168.30.1	255.255.255.0
PC4	192.168.40.101	192.168.40.1	255.255.255.0

第四步：验证 PC 之间的连通性，如表 1-40 所示。

表 1-40　验证 PC 之间的连通性

PC	端口	PC	端口	结果	原因
PC1	A：1/1	PC2	A：1/9	通	
PC1	A：1/1	VLAN 100	A：1/24	通	
PC1	A：1/1	VLAN 101	B：0/0/24	不通	
PC1	A：1/1	PC3	B：0/0/1	不通	

查看路由表，进一步分析上一步出现的现象和原因。
交换机 A：

```
DCRS-7604#show ip route
Total route items is 3, the matched route items is 3
Codes: C - connected, S - static, R - RIP derived, O - OSPF derived
       A - OSPF ASE, B - BGP derived, D - DVMRP derived
Destination         Mask              Nexthop       Interface     Preference
C   192.168.10.0    255.255.255.0     0.0.0.0       Vlan10        0
C   192.168.20.0    255.255.255.0     0.0.0.0       Vlan20        0
C   192.168.100.0   255.255.255.0     0.0.0.0       Vlan100       0
DCRS-7604#
```

交换机 B：

```
DCRS-5526S#show ip route
Total route items is 3, the matched route items is 3
Codes: C - connected, S - static, R - RIP derived, O - OSPF derived
       A - OSPF ASE, B - BGP derived, D - DVMRP derived
Destination         Mask              Nexthop       Interface     Preference
C   192.168.30.0    255.255.255.0     0.0.0.0       Vlan30        0
C   192.168.40.0    255.255.255.0     0.0.0.0       Vlan40        0
C   192.168.100.0   255.255.255.0     0.0.0.0       Vlan101       0
DCRS-5526S#
```

第五步：启动 RIP 协议，并将对应的直连网段配置到 RIP 进程中。
交换机 A：

```
DCRS-7604(Config)#router rip
DCRS-7604(Config-Router-Rip)#version 2
DCRS-7604(Config)#interface vlan 10
DCRS-7604(Config-If-Vlan10)#ip rip work
DCRS-7604(Config-If-Vlan10)#exit
DCRS-7604(Config)#interface vlan 20
DCRS-7604(Config-If-Vlan20)#ip rip work
DCRS-7604(Config-If-Vlan20)#exit
DCRS-7604(Config)#interface vlan 100
DCRS-7604(Config-If-Vlan100)#ip rip work
DCRS-7604(Config-If-Vlan100)#exit
```

验证配置：

```
DCRS-7604#show ip rip
RIP information:
Automatic network summarization is not in effect.
default metric for redistribute is :1.
neighbour is :NULL
```

preference is :120
RIP version information is :
interface   send version   receive version
Vlan 10   V2MC   V2
Vlan 20   V2MC   V2
Vlan 100   V2MC   V2
DCRS-7604#
DCRS-7604#show ip route
Total route items is 4, the matched route items is 4
Codes: C - connected, S - static, R - RIP derived, O - OSPF derived
       A - OSPF ASE, B - BGP derived, D - DVMRP derived

Destination	Mask	Nexthop	Interface	Preference
C 192.168.10.0	255.255.255.0	0.0.0.0	Vlan10	0
R 192.168.30.0	255.255.255.0	192.168.100.2	Vlan100	120
R 192.168.40.0	255.255.255.0	192.168.100.2	Vlan100	120
C 192.168.100.0	255.255.255.0	0.0.0.0	Vlan100	0

DCRS-7604#

R 表示 RIP 协议学习到的网段

交换机 B：

DCRS-5526S(Config)#router rip
DCRS-5526S(Config-Router-Rip)#version 2
DCRS-5526S(Config-Router-Rip)#exit
DCRS-5526S(Config)#interface vlan 30
DCRS-5526S(Config-If-Vlan30)#ip rip work
DCRS-5526S(Config-If-Vlan30)#exit
DCRS-5526S(Config)#interface vlan 40
DCRS-5526S(Config-If-Vlan40)#ip rip work
DCRS-5526S(Config-If-Vlan40)#exit
DCRS-5526S(Config)#interface vlan 101
DCRS-5526S(Config-If-Vlan101)#ip rip work
DCRS-5526S(Config-If-Vlan101)#exit
DCRS-5526S(Config)#

验证配置：

DCRS-5526S#show ip rip
RIP information:
Automatic network summarization is not in effect.
default metric for redistribute is :1.
neighbour is :NULL
preference is :120
RIP version information is :
interface   send version   receive version
Vlan 30   V2MC   V2

| Vlan 40 | V2MC | V2 |
| Vlan 101 | V2MC | V2 |

DCRS-5526S#
DCRS-5526S#show ip route
Total route items is 4, the matched route items is 4
Codes: C - connected, S - static, R - RIP derived, O - OSPF derived
   A - OSPF ASE, B - BGP derived, D - DVMRP derived

	Destination	Mask	Nexthop	Interface	Preference
R	192.168.10.0	255.255.255.0	192.168.100.1	Vlan101	120
C	192.168.30.0	255.255.255.0	0.0.0.0	Vlan30	0
C	192.168.40.0	255.255.255.0	0.0.0.0	Vlan40	0
C	192.168.100.0	255.255.255.0	0.0.0.0	Vlan101	0

DCRS-5526S#

思科设备的配置命令如下。

三层交换机 RS3560-A：

```
Switch>enable
Switch#configure terminal
Switch(config)#hostname RS3560-A
RS3560-A(config)#
RS3560-A(config)#vlan 10
RS3560-A(config-vlan)#ex
RS3560-A(config)#vlan 20
RS3560-A(config-vlan)#ex
RS3560-A(config)#vlan 100
RS3560-A(config-vlan)#ex
RS3560-A(config)#interface range fastEthernet 0/1-8
RS3560-A(config-if-range)#switchport access vlan 10
RS3560-A(config-if-range)#ex
RS3560-A(config)#interface range fastEthernet 0/9-16
RS3560-A(config-if-range)#switchport access vlan 20
RS3560-A(config-if-range)#ex
RS3560-A(config)#interface fastEthernet 0/24
RS3560-A(config-if)#switchport access vlan 100
RS3560-A(config-if)#ex
RS3560-A(config)#
RS3560-A(config)#interface vlan 10
RS3560-A(config-if)#
%LINK-5-CHANGED: Interface Vlan10, changed state to up
RS3560-A(config-if)#ip address 192.168.10.1 255.255.255.0
RS3560-A(config-if)#no shutdown
RS3560-A(config-if)#interface vlan 20
RS3560-A(config-if)#
%LINK-5-CHANGED: Interface Vlan20, changed state to up
RS3560-A(config-if)#ip address 192.168.20.1 255.255.255.0
```

RS3560-A(config-if)#no shutdown
RS3560-A(config-if)#interface vlan 100
RS3560-A(config-if)#
%LINK-5-CHANGED: Interface Vlan100, changed state to up
RS3560-A(config-if)#ip address 192.168.100.1 255.255.255.0
RS3560-A(config-if)#no shutdown
RS3560-A(config-if)#ex
RS3560-A(config)#ip routing
RS3560-A(config)#
RS3560-A#sh ip route
Codes: C - connected, S - static, I - IGRP, R - RIP, M - mobile, B - BGP
       D - EIGRP, EX - EIGRP external, O - OSPF, IA - OSPF inter area
       N1 - OSPF NSSA external type 1, N2 - OSPF NSSA external type 2
       E1 - OSPF external type 1, E2 - OSPF external type 2, E - EGP
       i - IS-IS, L1 - IS-IS level-1, L2 - IS-IS level-2, ia - IS-IS inter area
       * - candidate default, U - per-user static route, o - ODR
       P - periodic downloaded static route
Gateway of last resort is not set
C   192.168.10.0/24 is directly connected, Vlan10
C   192.168.20.0/24 is directly connected, Vlan20
C   192.168.100.0/24 is directly connected, Vlan100

三层交换机 RS3560-B：

Switch>enable
Switch#configure terminal
Switch(config)#hostname RS3560-B
RS3560-B(config)#vl
RS3560-B(config)#vlan 30
RS3560-B(config-vlan)#ex
RS3560-B(config)#vlan 40
RS3560-B(config-vlan)#ex
RS3560-B(config)#vlan 101
RS3560-B(config-vlan)#ex
RS3560-B(config)#interface range fastEthernet 0/1-8
RS3560-B(config-if-range)#switchport access vlan 30
RS3560-B(config-if-range)#interface range fastEthernet 0/9-16
RS3560-B(config-if-range)#switchport access vlan 40
RS3560-B(config-if-range)#ex
RS3560-B(config)#interface fastEthernet 0/24
RS3560-B(config-if)#switchport access vlan 101
RS3560-B(config)#interface vlan 30
RS3560-B(config-if)#
%LINK-5-CHANGED: Interface Vlan30, changed state to up
RS3560-B(config-if)#ip address 192.168.30.1 255.255.255.0
RS3560-B(config-if)#no shutdown

```
RS3560-B(config-if)#interface vlan 40
%LINK-5-CHANGED: Interface Vlan40, changed state to up
RS3560-B(config-if)#ip address 192.168.40.1 255.255.255.0
RS3560-B(config-if)#no shutdown
RS3560-B(config-if)#interface vlan 101
%LINK-5-CHANGED: Interface Vlan101, changed state to up
RS3560-B(config-if)#ip address 192.168.100.2 255.255.255.0
RS3560-B(config-if)#no shutdown
RS3560-B(config-if)#ex
RS3560-B(config)#ip routing
RS3560-B(config)#
RS3560-B#sh ip route
Codes: C - connected, S - static, I - IGRP, R - RIP, M - mobile, B - BGP
       D - EIGRP, EX - EIGRP external, O - OSPF, IA - OSPF inter area
       N1 - OSPF NSSA external type 1, N2 - OSPF NSSA external type 2
       E1 - OSPF external type 1, E2 - OSPF external type 2, E - EGP
       i - IS-IS, L1 - IS-IS level-1, L2 - IS-IS level-2, ia - IS-IS inter area
       * - candidate default, U - per-user static route, o - ODR
       P - periodic downloaded static route

Gateway of last resort is not set
    C    192.168.30.0/24 is directly connected, Vlan30
    C    192.168.40.0/24 is directly connected, Vlan40
    C    192.168.100.0/24 is directly connected, Vlan101
```

配置静态路由。

三层交换机 RS3560-A：

```
RS3560-A(config)#ip route 192.168.30.0 255.255.255.0 192.168.100.2
RS3560-A(config)#ip route 192.168.40.0 255.255.255.0 192.168.100.2
RS3560-A(config)#ex
RS3560-A#
%SYS-5-CONFIG_I: Configured from console by console
RS3560-A#sh ip route
Codes: C - connected, S - static, I - IGRP, R - RIP, M - mobile, B - BGP
       D - EIGRP, EX - EIGRP external, O - OSPF, IA - OSPF inter area
       N1 - OSPF NSSA external type 1, N2 - OSPF NSSA external type 2
       E1 - OSPF external type 1, E2 - OSPF external type 2, E - EGP
       i - IS-IS, L1 - IS-IS level-1, L2 - IS-IS level-2, ia - IS-IS inter area
       * - candidate default, U - per-user static route, o - ODR
       P - periodic downloaded static route

Gateway of last resort is not set
    C    192.168.10.0/24 is directly connected, Vlan10
    C    192.168.20.0/24 is directly connected, Vlan20
    S    192.168.30.0/24 [1/0] via 192.168.100.2
    S    192.168.40.0/24 [1/0] via 192.168.100.2
    C    192.168.100.0/24 is directly connected, Vlan100
```

配置 RIP 路由：

RS3560-A(config)#router rip
RS3560-A(config-router)#version 2
RS3560-A(config-router)#net
RS3560-A(config-router)#network 192.168.10.0
RS3560-A(config-router)#network 192.168.20.0
RS3560-A(config-router)#network 192.168.100.0
RS3560-A(config-router)#ex
RS3560-A(config)#ex
RS3560-A#
RS3560-A#sh ip route
Codes: C - connected, S - static, I - IGRP, R - RIP, M - mobile, B - BGP
       D - EIGRP, EX - EIGRP external, O - OSPF, IA - OSPF inter area
       N1 - OSPF NSSA external type 1, N2 - OSPF NSSA external type 2
       E1 - OSPF external type 1, E2 - OSPF external type 2, E - EGP
       i - IS-IS, L1 - IS-IS level-1, L2 - IS-IS level-2, ia - IS-IS inter area
       * - candidate default, U - per-user static route, o - ODR
       P - periodic downloaded static route

Gateway of last resort is not set

C   192.168.10.0/24 is directly connected, Vlan10
C   192.168.20.0/24 is directly connected, Vlan20
R   192.168.30.0/24 [120/1] via 192.168.100.2, 00:00:10, Vlan100
R   192.168.40.0/24 [120/1] via 192.168.100.2, 00:00:10, Vlan100
C   192.168.100.0/24 is directly connected, Vlan100
RS3560-A#sh run
Building configuration...
Current configuration : 1863 bytes
!
version 12.2
no service timestamps log datetime msec
no service timestamps debug datetime msec
no service password-encryption
!
hostname RS3560-A
!
!
ip routing
!
!
!
spanning-tree mode pvst
!
!
!
interface FastEthernet0/1
 switchport access vlan 10

!
interface FastEthernet0/2
　switchport access vlan 10
!
interface FastEthernet0/3
　switchport access vlan 10
!
interface FastEthernet0/4
　switchport access vlan 10
!
interface FastEthernet0/5
　switchport access vlan 10
!
interface FastEthernet0/6
　switchport access vlan 10
!
interface FastEthernet0/7
　switchport access vlan 10
!
interface FastEthernet0/8
　switchport access vlan 10
!
interface FastEthernet0/9
　switchport access vlan 20
!
interface FastEthernet0/10
　switchport access vlan 20
!
interface FastEthernet0/11
　switchport access vlan 20
!
interface FastEthernet0/12
　switchport access vlan 20
!
interface FastEthernet0/13
　switchport access vlan 20
!
interface FastEthernet0/14
　switchport access vlan 20
!
interface FastEthernet0/15
　switchport access vlan 20
!
interface FastEthernet0/16
　switchport access vlan 20
!
interface FastEthernet0/17

```
!
interface FastEthernet0/18
!
interface FastEthernet0/19
!
interface FastEthernet0/20
!
interface FastEthernet0/21
!
interface FastEthernet0/22
!
interface FastEthernet0/23
!
interface FastEthernet0/24
 switchport access vlan 100
!
interface GigabitEthernet0/1
!
interface GigabitEthernet0/2
!
interface Vlan1
 no ip address
 shutdown
!
interface Vlan10
 ip address 192.168.10.1 255.255.255.0
!
interface Vlan20
 ip address 192.168.20.1 255.255.255.0
!
interface Vlan100
 ip address 192.168.100.1 255.255.255.0
!
router rip
 version 2
 network 192.168.10.0
 network 192.168.20.0
 network 192.168.100.0
!
ip classless
!
ip flow-export version 9
!
!
!
line con 0
```

!
line aux 0
!
line vty 0 4
 login
!
!
!
end

RS3560-A#
RS3560-B(config)#router rip
RS3560-B(config-router)#ver
RS3560-B(config-router)#version 2
RS3560-B(config-router)#net
RS3560-B(config-router)#network 192.168.30.0
RS3560-B(config-router)#network 192.168.40.0
RS3560-B(config-router)#network 192.168.100.0
RS3560-B(config-router)#^Z
RS3560-B#
RS3560-B#sh ip route
Codes: C - connected, S - static, I - IGRP, R - RIP, M - mobile, B - BGP
       D - EIGRP, EX - EIGRP external, O - OSPF, IA - OSPF inter area
       N1 - OSPF NSSA external type 1, N2 - OSPF NSSA external type 2
       E1 - OSPF external type 1, E2 - OSPF external type 2, E - EGP
       i - IS-IS, L1 - IS-IS level-1, L2 - IS-IS level-2, ia - IS-IS inter area
       * - candidate default, U - per-user static route, o - ODR
       P - periodic downloaded static route

Gateway of last resort is not set
    R   192.168.10.0/24 [120/1] via 192.168.100.1, 00:00:03, Vlan101
    R   192.168.20.0/24 [120/1] via 192.168.100.1, 00:00:03, Vlan101
    C   192.168.30.0/24 is directly connected, Vlan30
    C   192.168.40.0/24 is directly connected, Vlan40
    C   192.168.100.0/24 is directly connected, Vlan101
RS3560-B#sh run
Building configuration...
Current configuration : 1863 bytes
!
version 12.2
no service timestamps log datetime msec
no service timestamps debug datetime msec
no service password-encryption
!
hostname RS3560-B
!
ip routing
!

```
spanning-tree mode pvst
!
interface FastEthernet0/1
  switchport access vlan 30
!
interface FastEthernet0/2
  switchport access vlan 30
!
interface FastEthernet0/3
  switchport access vlan 30
!
interface FastEthernet0/4
  switchport access vlan 30
!
interface FastEthernet0/5
  switchport access vlan 30
!
interface FastEthernet0/6
  switchport access vlan 30
!
interface FastEthernet0/7
  switchport access vlan 30
!
interface FastEthernet0/8
  switchport access vlan 30
!
interface FastEthernet0/9
  switchport access vlan 40
!
interface FastEthernet0/10
  switchport access vlan 40
!
interface FastEthernet0/11
  switchport access vlan 40
!
interface FastEthernet0/12
  switchport access vlan 40
!
interface FastEthernet0/13
  switchport access vlan 40
!
interface FastEthernet0/14
  switchport access vlan 40
!
interface FastEthernet0/15
  switchport access vlan 40
!
```

```
interface FastEthernet0/16
  switchport access vlan 40
!
interface FastEthernet0/17
!
interface FastEthernet0/18
!
interface FastEthernet0/19
!
interface FastEthernet0/20
!
interface FastEthernet0/21
!
interface FastEthernet0/22
!
interface FastEthernet0/23
!
interface FastEthernet0/24
  switchport access vlan 101
!
interface GigabitEthernet0/1
!
interface GigabitEthernet0/2
!
interface Vlan1
  no ip address
  shutdown
!
interface Vlan30
  ip address 192.168.30.1 255.255.255.0
!
interface Vlan40
  ip address 192.168.40.1 255.255.255.0
!
interface Vlan101
  ip address 192.168.100.2 255.255.255.0
!
router rip
  version 2
  network 192.168.30.0
  network 192.168.40.0
  network 192.168.100.0
!
ip classless
!
ip flow-export version 9
!
```

```
!
line con 0
!
line aux 0
!
line vty 0 4
 login
!
!
end
RS3560-B#
```

第六步：验证 PC 之间的连通性，如表 1-41 所示。

表 1-41 验证 PC 之间的连通性

PC	端口	PC	端口	结果	原因
PC1	A: 1/1	PC2	A: 1/9	通	
PC1	A: 1/1	VLAN 100	A: 1/24	通	
PC1	A: 1/1	VLAN 101	B: 0/0/24	通	
PC1	A: 1/1	PC3	B: 0/0/1	通	

七、注意事项和排错

1. 全局启动"router rip"之后，交换机自动会在所有的虚接口上启动 RIP 协议。
2. 可以在单个虚接口上禁止 RIP 协议。

八、思考题

1. 如果在交换机 A 的 VLAN 100 上禁用 RIP 协议，PC1 是否还能 ping 通 PC3？
2. 如果在交换机 B 的 VLAN 30 上禁用 RIP 协议，PC1 是否还能 ping 通 PC3？

实训项目六 多层交换机动态 RIP 路由配置

一、实训目的

1. 学会在多层交换机上配置动态 RIP 路由，实现不同 VLAN 或子网之间的自动通信。
2. 理解动态路由和静态路由的区别。
3. 验证动态 RIP 路由配置，诊断并解决路由配置中的问题。

二、实训设备

(1) DCS-3926S 交换机 2 台。
(2) DCRS-5526S 交换机 2 台。

(3) PC 机 4 台。
(4) 直通网线 7 根。

三、实训拓扑

该实验拓扑结构如图 1-43 所示。

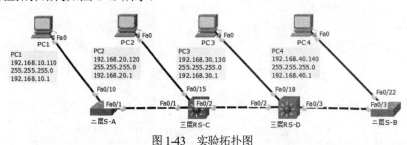

图 1-43　实验拓扑图

四、实训要求

1. 参照如图 1-43 所示进行连线。
2. 实训结果：PC1---ping---PC2----PC3----PC4　全通。

五、实训步骤

第一步：交换机 A、B、C、D 的配置如下。
A 划分 vl 10(9~12)、vl 20(13~16)，Trunk (1)。
B 划分 vl 30(17~20)、vl 40(21~24)，Trunk (3)。
C 划分 vl 10(9~12)、vl 20(13~16)、vl 100(2)，Trunk (1)。
D 划分 vl 30(17~20)、vl 40(21~24)、vl 101(2)，Trunk (3)。
第二步：设置三层交换机 C 启用三层路由功能。
设置 vl 10、vl 20、vl 100 的接口地址：

　int vl 10　192.168.10.1/24；
　int vl 20　192.168.20.1/24；
　int vl 100 192.168.100.1/30。

第三步：设置三层交换机 D 启用三层路由功能；
设置 vl 30、vl 40、vl 101 的接口地址：

　int vl 30　192.168.30.1/24 ；
　int vl 40　192.168.40.1/24；
　int vl 101 192.168.100.2/30。

方法 1。配置交换机 C 的动态路由 RIP：

C(config)#router rip

```
C(config-Router)#version 2
C(config)#int vl 10
C(config-if-vlan10)#ip rip work
C(config)#int vl 20
C(config-if-vlan20)#ip rip work
C(config)#int vl 100
C(config-if-vlan100)#ip rip work
```

配置交换机 D 的动态路由 RIP：

```
D(config)#router rip
D(config-Router)#version 2
D(config)#int vl 30
D(config-if-vlan30)#ip rip work
D(config)#int vl 40
D(config-if-vlan40)#ip rip work
D(config)#int vl 101
D(config-if-vlan101)#ip rip work
```

验证：show ip rip   show ip route

方法 2。

配置交换机 C 的动态路由 RIP。

```
C(config)#router rip
C(config-Router)#version 2
C(config-Router)#network vl 10
C(config-Router)#network vl 20
C(config-Router)#network vl 100
```

配置交换机 D 的动态路由 RIP：

```
D(config)#router rip
D(config-Router)#version 2
D(config-Router)#network vl 30
D(config-Router)#network vl 40
D(config-Router)#network vl 101
```

验证：show ip rip   show ip route

思科设备采取方法 3。

方法 3。

配置交换机 C 的动态路由 RIP：

```
RS3560-C(config)#router rip
```

RS3560-C(config-router)#version 2
RS3560-C(config-router)#network 192.168.10.0
RS3560-C(config-router)#network 192.168.20.0
RS3560-C(config-router)#network 192.168.100.0

配置交换机 D 的动态路由 RIP：

RS3560-D(config)#router rip
RS3560-D(config-router)#version 2
RS3560-D(config-router)#network 192.168.30.0
RS3560-D(config-router)#network 192.168.40.0
RS3560-D(config-router)#network 192.168.100.0

验证：show ip rip   show ip route

第四步：根据连线位置正确配置 PC1、PC2、PC3、PC4 的地址(注意配置网关地址以及与 VLAN 的对应关系)。

实验结果：PC1---ping---PC2----PC3----PC4   全通。
查看交换机状态：show vlan   show run   show ip route   show ip rip。

实验十七 三层交换机 OSPF 动态路由

一、实验目的

1. 掌握三层交换机之间通过 OSPF 协议实现网段互通的配置方法。
2. 理解 RIP 协议和 OSPF 协议内部实现的不同点。

二、相关知识

OSPF 是一种链路状态路由协议，它通过收集网络中所有路由器的链路状态信息来计算最短路径。OSPF 支持无类域间路由(CIDR)和可变长度子网掩码(VLSM)，能够更有效地利用 IP 地址空间。当两台三层交换机级联时，为了保证每台交换机上连接的网段可以和另一台交换机上连接的网段互通，可以在三层交换机上配置 OSPF 路由协议，以实现动态路由。

三层交换机 OSPF 动态路由的应用场景如下。

(1) 校园网络：在校园网络中，不同部门可能需要访问不同的资源，OSPF 动态路由可以实现资源的隔离和访问控制，同时支持网络的扩展和变化。

(2) 企业网络管理：在企业网络中，OSPF 动态路由可以实现网络的自动管理和控制，提高网络的可扩展性和灵活性。

(3) 数据中心网络：数据中心中的服务器和存储设备可能分布在不同的 VLAN 中，OSPF 动态路由可以实现这些设备之间的通信，同时保持网络的高性能。

三、实验设备

1. DCRS-7604(或 6804)交换机 1 台。
2. DCRS-5526S 交换机 1 台。
3. PC 机 2~4 台。
4. Console 线 1~2 根。
5. 直通网线 2~4 根。

四、实验拓扑

该实验拓扑结构如图 1-44 所示。

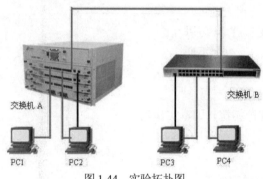

图 1-44 实验拓扑图

五、实验要求

1. 在交换机 A 和交换机 B 上分别划分基于端口的 VLAN，如表 1-42 所示。

表 1-42 交换机 A 和交换机 B 的 VLAN 划分

交换机	VLAN	端口成员
交换机 A	10	1~8
	20	9~16
	100	24
交换机 B	30	1~8
	40	9~16
	101	24

2. 交换机 A 和交换机 B 通过 24 口进行级联。
3. 配置交换机 A 和交换机 B 各 VLAN 虚拟接口的 IP 地址，如表 1-43 所示。

表 1-43 交换机 A 和交换机 B 的 IP 配置

VLAN 10	VLAN 20	VLAN 30	VLAN 40	VLAN 100	VLAN 101
192.168.10.1	192.168.20.1	192.168.30.1	192.168.40.1	192.168.100.1	192.168.100.2

4. 参照如表 1-44 所示配置 PC 的 IP 地址。

表 1-44　PC 的配置

设备	IP 地址	gateway	Mask
PC1	192.168.10.101	192.168.10.1	255.255.255.0
PC2	192.168.20.101	192.168.20.1	255.255.255.0
PC3	192.168.30.101	192.168.30.1	255.255.255.0
PC4	192.168.40.101	192.168.40.1	255.255.255.0

验证结果注意下列问题。

(1) 没有 OSPF 路由协议之前：

　　PC1、PC2 与 PC3、PC4 可以互通。

　　PC1、PC2 与 PC3、PC4 不通。

(2) 配置 RIP 路由协议之后：

　　4 台 PC 机之间都可以互通。

　　若实验结果和理论相符，则本实验完成。

六、实验步骤

神州数码设备的配置步骤如下。

第一步：交换机全部恢复出厂设置，配置交换机的 VLAN 信息。

交换机 A：

```
DCRS-7604#conf
DCRS-7604(Config)#vlan 10
DCRS-7604(Config-Vlan10)#switchport interface ethernet 1/1-8
Set the port Ethernet1/1 access vlan 10 successfully
Set the port Ethernet1/2 access vlan 10 successfully
Set the port Ethernet1/3 access vlan 10 successfully
Set the port Ethernet1/4 access vlan 10 successfully
Set the port Ethernet1/5 access vlan 10 successfully
Set the port Ethernet1/6 access vlan 10 successfully
Set the port Ethernet1/7 access vlan 10 successfully
Set the port Ethernet1/8 access vlan 10 successfully
DCRS-7604(Config-Vlan10)#exit
DCRS-7604(Config)#vlan 20
DCRS-7604(Config-Vlan20)#switchport interface ethernet 1/9-16
Set the port Ethernet1/9 access vlan 20 successfully
Set the port Ethernet1/10 access vlan 20 successfully
Set the port Ethernet1/11 access vlan 20 successfully
Set the port Ethernet1/12 access vlan 20 successfully
Set the port Ethernet1/13 access vlan 20 successfully
Set the port Ethernet1/14 access vlan 20 successfully
```

Set the port Ethernet1/15 access vlan 20 successfully
Set the port Ethernet1/16 access vlan 20 successfully
DCRS-7604(Config-Vlan20)#exit
DCRS-7604(Config)#vlan 100
DCRS-7604(Config-Vlan100)#switchport interface ethernet 1/24
Set the port Ethernet1/24 access vlan 100 successfully
DCRS-7604(Config-Vlan100)#exit
DCRS-7604(Config)#

验证配置：

DCRS-7604#show vlan

VLAN	Name	Type	Media	Ports	
1	default	Static	ENET	Ethernet1/17	Ethernet1/18
				Ethernet1/19	Ethernet1/20
				Ethernet1/21	Ethernet1/22
				Ethernet1/23	Ethernet1/25
				Ethernet1/26	Ethernet1/27
				Ethernet1/28	
10	VLAN0010	Static	ENET	Ethernet1/1	Ethernet1/2
				Ethernet1/3	Ethernet1/4
				Ethernet1/5	Ethernet1/6
				Ethernet1/7	Ethernet1/8
20	VLAN0020	Static	ENET	Ethernet1/9	Ethernet1/10
				Ethernet1/11	Ethernet1/12
				Ethernet1/13	Ethernet1/14
				Ethernet1/15	Ethernet1/16
100	VLAN0100	Static	ENET	Ethernet1/24	

DCRS-7604#

交换机 B：

DCRS-5526S(Config)#vlan 30
DCRS-5526S(Config-Vlan30)#switchport interface ethernet 0/0/1-8
Set the port Ethernet0/0/1 access vlan 30 successfully
Set the port Ethernet0/0/2 access vlan 30 successfully
Set the port Ethernet0/0/3 access vlan 30 successfully
Set the port Ethernet0/0/4 access vlan 30 successfully
Set the port Ethernet0/0/5 access vlan 30 successfully
Set the port Ethernet0/0/6 access vlan 30 successfully
Set the port Ethernet0/0/7 access vlan 30 successfully
Set the port Ethernet0/0/8 access vlan 30 successfully
DCRS-5526S(Config-Vlan30)#exit
DCRS-5526S(Config)#vlan 40
DCRS-5526S(Config-Vlan40)#switchport interface ethernet 0/0/9-16
Set the port Ethernet0/0/9 access vlan 40 successfully

Set the port Ethernet0/0/10 access vlan 40 successfully
Set the port Ethernet0/0/11 access vlan 40 successfully
Set the port Ethernet0/0/12 access vlan 40 successfully
Set the port Ethernet0/0/13 access vlan 40 successfully
Set the port Ethernet0/0/14 access vlan 40 successfully
Set the port Ethernet0/0/15 access vlan 40 successfully
Set the port Ethernet0/0/16 access vlan 40 successfully
DCRS-5526S(Config-Vlan40)#exit
DCRS-5526S(Config)#vlan 101
DCRS-5526S(Config-Vlan101)#switchport interface ethernet 0/0/24
Set the port Ethernet0/0/24 access vlan 101 successfully
DCRS-5526S(Config-Vlan101)#exit
DCRS-5526S(Config)#

验证配置：

DCRS-5526S#show vlan

VLAN	Name	Type	Media	Ports	
1	default	Static	ENET	Ethernet0/0/17	Ethernet0/0/18
				Ethernet0/0/19	Ethernet0/0/20
				Ethernet0/0/21	Ethernet0/0/22
				Ethernet0/0/23	
30	VLAN0030	Static	ENET	Ethernet0/0/1	Ethernet0/0/2
				Ethernet0/0/3	Ethernet0/0/4
				Ethernet0/0/5	Ethernet0/0/6
				Ethernet0/0/7	Ethernet0/0/8
40	VLAN0040	Static	ENET	Ethernet0/0/9	Ethernet0/0/10
				Ethernet0/0/11	Ethernet0/0/12
				Ethernet0/0/13	Ethernet0/0/14
				Ethernet0/0/15	Ethernet0/0/16
101	VLAN0101	Static	ENET	Ethernet0/0/24	

DCRS-5526S#

第二步：配置交换机各 VLAN 虚接口的 IP 地址。

交换机 A：

DCRS-7604(Config)#int vlan 10
DCRS-7604(Config-If-Vlan10)#ip address 192.168.10.1 255.255.255.0
DCRS-7604(Config-If-Vlan10)#no shut
DCRS-7604(Config-If-Vlan10)#exit
DCRS-7604(Config)#int vlan 20
DCRS-7604(Config-If-Vlan20)#ip address 192.168.20.1 255.255.255.0
DCRS-7604(Config-If-Vlan20)#no shut
DCRS-7604(Config-If-Vlan20)#exit
DCRS-7604(Config)#int vlan 100
DCRS-7604(Config-If-Vlan100)#ip address 192.168.100.1 255.255.255.0

第1章 交换机实验

```
DCRS-7604(Config-If-Vlan100)#no shut
DCRS-7604(Config-If-Vlan100)#
DCRS-7604(Config-If-Vlan100)#exit
DCRS-7604(Config)#
```

交换机 B：

```
DCRS-5526S(Config)#int vlan 30
DCRS-5526S(Config-If-Vlan30)#ip address 192.168.30.1 255.255.255.0
DCRS-5526S(Config-If-Vlan30)#no shut
DCRS-5526S(Config-If-Vlan30)#exit
DCRS-5526S(Config)#interface vlan 40
DCRS-5526S(Config-If-Vlan40)#ip address 192.168.40.1 255.255.255.0
DCRS-5526S(Config-If-Vlan40)#exit
DCRS-5526S(Config)#int vlan 101
DCRS-5526S(Config-If-Vlan101)#ip address 192.168.100.2 255.255.255.0
DCRS-5526S(Config-If-Vlan101)#exit
DCRS-5526S(Config)#
```

第三步：参照表 1-45 所示配置 PC 的 IP 地址(注意配置网关)。

表 1-45　PC 的配置

设备	IP 地址	gateway	Mask
PC1	192.168.10.101	192.168.10.1	255.255.255.0
PC2	192.168.20.101	192.168.20.1	255.255.255.0
PC3	192.168.30.101	192.168.30.1	255.255.255.0
PC4	192.168.40.101	192.168.40.1	255.255.255.0

第四步：验证 PC 之间的连通性，如表 1-46 所示。

表 1-46　验证 PC 之间的连通性

PC	端口	PC	端口	结果	原因
PC1	A：1/1	PC2	A：1/9	通	
PC1	A：1/1	VLAN 100	A：1/24	通	
PC1	A：1/1	VLAN 101	B：0/0/24	不通	
PC1	A：1/1	PC3	B：0/0/1	不通	

查看路由表，进一步分析上一步操作导致的结果和原因。
交换机 A：

```
DCRS-7604#show ip route
Total route items is 3, the matched route items is 3
Codes: C - connected, S - static, R - RIP derived, O - OSPF derived
```

```
      A - OSPF ASE, B - BGP derived, D - DVMRP derived
Destination        Mask              Nexthop    Interface    Preference
C   192.168.10.0    255.255.255.0    0.0.0.0    Vlan10       0
C   192.168.20.0    255.255.255.0    0.0.0.0    Vlan20       0
C   192.168.100.0   255.255.255.0    0.0.0.0    Vlan100      0
DCRS-7604#
```

交换机 B：

```
DCRS-5526S#show ip route
Total route items is 3, the matched route items is 3
Codes: C - connected, S - static, R - RIP derived, O - OSPF derived
      A - OSPF ASE, B - BGP derived, D - DVMRP derived
Destination        Mask              Nexthop    Interface    Preference
C   192.168.30.0    255.255.255.0    0.0.0.0    Vlan30       0
C   192.168.40.0    255.255.255.0    0.0.0.0    Vlan40       0
C   192.168.100.0   255.255.255.0    0.0.0.0    Vlan101      0
DCRS-5526S#
```

第五步：启动 OSPF 协议，并将对应的直连网段配置到 OSPF 进程中。

交换机 A：

```
DCRS-7604(Config)#router ospf
OSPF protocol is working, please waiting.......
OSPF protocol has enabled!
DCRS-7604(Config-Router-Ospf)#exit
DCRS-7604(Config)#interface vlan 10
DCRS-7604(Config-If-Vlan10)#ip ospf enable area 0
DCRS-7604(Config-If-Vlan10)#
DCRS-7604(Config)#interface vlan 20
DCRS-7604(Config-If-Vlan20)#ip ospf enable area 0
DCRS-7604(Config-If-Vlan20)#exit
DCRS-7604(Config)#interface vlan 100
DCRS-7604(Config-If-Vlan100)#
DCRS-7604(Config-If-Vlan100)#ip ospf enable area 0
DCRS-7604(Config-If-Vlan100)#exit
DCRS-7604(Config)#
```

验证配置：

```
DCRS-7604#show ip route
Total route items is 4, the matched route items is 4
Codes: C - connected, S - static, R - RIP derived, O - OSPF derived
      A - OSPF ASE, B - BGP derived, D - DVMRP derived
Destination        Mask              Nexthop          Interface    Preference
C   192.168.10.0    255.255.255.0    0.0.0.0          Vlan10       0
O   192.168.30.0    255.255.255.0    192.168.100.2    Vlan100      110
O   192.168.40.0    255.255.255.0    192.168.100.2    Vlan100      110
```

| C | 192.168.100.0 | 255.255.255.0 | 0.0.0.0 | Vlan100 | 0 |

DCRS-7604#

交换机 B(5526S 的配置界面和 7604 是一样的)。

思科设备的配置命令如下。

三层交换机 RS3560-A：

Switch>enable
Switch#configure terminal
Switch(config)#hostname RS3560-A
RS3560-A(config)#
RS3560-A(config)#vlan 10
RS3560-A(config-vlan)#ex
RS3560-A(config)#vlan 20
RS3560-A(config-vlan)#ex
RS3560-A(config)#vlan 100
RS3560-A(config-vlan)#ex
RS3560-A(config)#interface range fastEthernet 0/1-8
RS3560-A(config-if-range)#switchport access vlan 10
RS3560-A(config-if-range)#ex
RS3560-A(config)#interface range fastEthernet 0/9-16
RS3560-A(config-if-range)#switchport access vlan 20
RS3560-A(config-if-range)#ex
RS3560-A(config)#interface fastEthernet 0/24
RS3560-A(config-if)#switchport access vlan 100
RS3560-A(config-if)#ex
RS3560-A(config)#
RS3560-A(config)#interface vlan 10
RS3560-A(config-if)#
%LINK-5-CHANGED: Interface Vlan10, changed state to up
RS3560-A(config-if)#ip address 192.168.10.1 255.255.255.0
RS3560-A(config-if)#no shutdown
RS3560-A(config-if)#interface vlan 20
RS3560-A(config-if)#
%LINK-5-CHANGED: Interface Vlan20, changed state to up
RS3560-A(config-if)#ip address 192.168.20.1 255.255.255.0
RS3560-A(config-if)#no shutdown
RS3560-A(config-if)#interface vlan 100
RS3560-A(config-if)#
%LINK-5-CHANGED: Interface Vlan100, changed state to up
RS3560-A(config-if)#ip address 192.168.100.1 255.255.255.0
RS3560-A(config-if)#no shutdown
RS3560-A(config-if)#ex
RS3560-A(config)#ip routing
RS3560-A(config)#
RS3560-A#sh ip route

```
Codes: C - connected, S - static, I - IGRP, R - RIP, M - mobile, B - BGP
       D - EIGRP, EX - EIGRP external, O - OSPF, IA - OSPF inter area
       N1 - OSPF NSSA external type 1, N2 - OSPF NSSA external type 2
       E1 - OSPF external type 1, E2 - OSPF external type 2, E - EGP
       i - IS-IS, L1 - IS-IS level-1, L2 - IS-IS level-2, ia - IS-IS inter area
       * - candidate default, U - per-user static route, o - ODR
       P - periodic downloaded static route
Gateway of last resort is not set
     C    192.168.10.0/24 is directly connected, Vlan10
     C    192.168.20.0/24 is directly connected, Vlan20
     C    192.168.100.0/24 is directly connected, Vlan100
```

三层交换机 RS3560-B：

```
Switch>enable
Switch#configure terminal
Switch(config)#hostname RS3560-B
RS3560-B(config)#vl
RS3560-B(config)#vlan 30
RS3560-B(config-vlan)#ex
RS3560-B(config)#vlan 40
RS3560-B(config-vlan)#ex
RS3560-B(config)#vlan 101
RS3560-B(config-vlan)#ex
RS3560-B(config)#interface range fastEthernet 0/1-8
RS3560-B(config-if-range)#switchport access vlan 30
RS3560-B(config-if-range)#interface range fastEthernet 0/9-16
RS3560-B(config-if-range)#switchport access vlan 40
RS3560-B(config-if-range)#ex
RS3560-B(config)#interface fastEthernet 0/24
RS3560-B(config-if)#switchport access vlan 101
RS3560-B(config)#interface vlan 30
RS3560-B(config-if)#
%LINK-5-CHANGED: Interface Vlan30, changed state to up
RS3560-B(config-if)#ip address 192.168.30.1 255.255.255.0
RS3560-B(config-if)#no shutdown
RS3560-B(config-if)#interface vlan 40
%LINK-5-CHANGED: Interface Vlan40, changed state to up
RS3560-B(config-if)#ip address 192.168.40.1 255.255.255.0
RS3560-B(config-if)#no shutdown
RS3560-B(config-if)#interface vlan 101
%LINK-5-CHANGED: Interface Vlan101, changed state to up
RS3560-B(config-if)#ip address 192.168.100.2 255.255.255.0
RS3560-B(config-if)#no shutdown
RS3560-B(config-if)#ex
RS3560-B(config)#ip routing
RS3560-B(config)#
```

RS3560-B#sh ip route
Codes: C - connected, S - static, I - IGRP, R - RIP, M - mobile, B - BGP
       D - EIGRP, EX - EIGRP external, O - OSPF, IA - OSPF inter area
       N1 - OSPF NSSA external type 1, N2 - OSPF NSSA external type 2
       E1 - OSPF external type 1, E2 - OSPF external type 2, E - EGP
       i - IS-IS, L1 - IS-IS level-1, L2 - IS-IS level-2, ia - IS-IS inter area
       * - candidate default, U - per-user static route, o - ODR
       P - periodic downloaded static route
Gateway of last resort is not set
C   192.168.30.0/24 is directly connected, Vlan30
C   192.168.40.0/24 is directly connected, Vlan40
C   192.168.100.0/24 is directly connected, Vlan101

配置 OSPF 路由。
三层交换机 RS3560-A：

RS3560-A(config)#router ospf 1
RS3560-A(config-router)#network 192.168.100.0 255.255.255.0 area 0
RS3560-A(config-router)#network 192.168.10.0 255.255.255.0 area 0
RS3560-A(config-router)#network 192.168.20.0 255.255.255.0 area 0
RS3560-A(config-router)#^Z
RS3560-A#
RS3560-A#sh ip route
Codes: C - connected, S - static, I - IGRP, R - RIP, M - mobile, B - BGP
       D - EIGRP, EX - EIGRP external, O - OSPF, IA - OSPF inter area
       N1 - OSPF NSSA external type 1, N2 - OSPF NSSA external type 2
       E1 - OSPF external type 1, E2 - OSPF external type 2, E - EGP
       i - IS-IS, L1 - IS-IS level-1, L2 - IS-IS level-2, ia - IS-IS inter area
       * - candidate default, U - per-user static route, o - ODR
       P - periodic downloaded static route
Gateway of last resort is not set
C   192.168.10.0/24 is directly connected, Vlan10
C   192.168.20.0/24 is directly connected, Vlan20
O   192.168.30.0/24 [110/2] via 192.168.100.2, 00:00:14, Vlan100
O   192.168.40.0/24 [110/2] via 192.168.100.2, 00:00:14, Vlan100
C   192.168.100.0/24 is directly connected, Vlan100
RS3560-A#sh run
Building configuration...
Current configuration : 1929 bytes
!
version 12.2
no service timestamps log datetime msec
no service timestamps debug datetime msec
no service password-encryption
!
hostname RS3560-A
!

```
ip routing
!
spanning-tree mode pvst
!
interface FastEthernet0/1
  switchport access vlan 10
!
interface FastEthernet0/2
  switchport access vlan 10
!
interface FastEthernet0/3
  switchport access vlan 10
!
interface FastEthernet0/4
  switchport access vlan 10
!
interface FastEthernet0/5
  switchport access vlan 10
!
interface FastEthernet0/6
  switchport access vlan 10
!
interface FastEthernet0/7
  switchport access vlan 10
!
interface FastEthernet0/8
  switchport access vlan 10
!
interface FastEthernet0/9
  switchport access vlan 20
!
interface FastEthernet0/10
  switchport access vlan 20
!
interface FastEthernet0/11
  switchport access vlan 20
!
interface FastEthernet0/12
  switchport access vlan 20
!
interface FastEthernet0/13
  switchport access vlan 20
!
interface FastEthernet0/14
  switchport access vlan 20
!
interface FastEthernet0/15
```

```
  switchport access vlan 20
!
interface FastEthernet0/16
  switchport access vlan 20
!
interface FastEthernet0/17
!
interface FastEthernet0/18
!
interface FastEthernet0/19
!
interface FastEthernet0/20
!
interface FastEthernet0/21
!
interface FastEthernet0/22
!
interface FastEthernet0/23
!
interface FastEthernet0/24
  switchport access vlan 100
!
interface GigabitEthernet0/1
!
interface GigabitEthernet0/2
!
interface Vlan1
  no ip address
  shutdown
!
interface Vlan10
  ip address 192.168.10.1 255.255.255.0
!
interface Vlan20
  ip address 192.168.20.1 255.255.255.0
!
interface Vlan100
  ip address 192.168.100.1 255.255.255.0
!
router ospf 1
  log-adjacency-changes
  network 192.168.100.0 0.0.0.255 area 0
  network 192.168.10.0 0.0.0.255 area 0
  network 192.168.20.0 0.0.0.255 area 0
!
ip classless
!
```

```
ip flow-export version 9
!
line con 0
!
line aux 0
!
line vty 0 4
 login
!
end
RS3560-A#
```

三层交换机 RS3560-B：

```
RS3560-B(config)#router ospf 2
RS3560-B(config-router)#network 192.168.100.0 255.255.255.0 area 0
RS3560-B(config-router)#network 192.168.30.0 255.255.255.0 area 0
RS3560-B(config-router)#network 192.168.40.0 255.255.255.0 area 0
RS3560-B(config-router)#ex
RS3560-B#
RS3560-B#sh ip route
Codes: C - connected, S - static, I - IGRP, R - RIP, M - mobile, B - BGP
       D - EIGRP, EX - EIGRP external, O - OSPF, IA - OSPF inter area
       N1 - OSPF NSSA external type 1, N2 - OSPF NSSA external type 2
       E1 - OSPF external type 1, E2 - OSPF external type 2, E - EGP
       i - IS-IS, L1 - IS-IS level-1, L2 - IS-IS level-2, ia - IS-IS inter area
       * - candidate default, U - per-user static route, o - ODR
       P - periodic downloaded static route

Gateway of last resort is not set
O    192.168.10.0/24 [110/2] via 192.168.100.1, 00:00:11, Vlan101
O    192.168.20.0/24 [110/2] via 192.168.100.1, 00:00:11, Vlan101
C    192.168.30.0/24 is directly connected, Vlan30
C    192.168.40.0/24 is directly connected, Vlan40
C    192.168.100.0/24 is directly connected, Vlan101
RS3560-B#sh run
Building configuration...
Current configuration : 1929 bytes
!
version 12.2
no service timestamps log datetime msec
no service timestamps debug datetime msec
no service password-encryption
!
hostname RS3560-B
!
ip routing
!
spanning-tree mode pvst
```

!
interface FastEthernet0/1
　switchport access vlan 30
!
interface FastEthernet0/2
　switchport access vlan 30
!
interface FastEthernet0/3
　switchport access vlan 30
!
interface FastEthernet0/4
　switchport access vlan 30
!
interface FastEthernet0/5
　switchport access vlan 30
!
interface FastEthernet0/6
　switchport access vlan 30
!
interface FastEthernet0/7
　switchport access vlan 30
!
interface FastEthernet0/8
　switchport access vlan 30
!
interface FastEthernet0/9
　switchport access vlan 40
!
interface FastEthernet0/10
　switchport access vlan 40
!
interface FastEthernet0/11
　switchport access vlan 40
!
interface FastEthernet0/12
　switchport access vlan 40
!
interface FastEthernet0/13
　switchport access vlan 40
!
interface FastEthernet0/14
　switchport access vlan 40
!
interface FastEthernet0/15
　switchport access vlan 40
!
interface FastEthernet0/16

```
  switchport access vlan 40
!
interface FastEthernet0/17
!
interface FastEthernet0/18
!
interface FastEthernet0/19
!
interface FastEthernet0/20
!
interface FastEthernet0/21
!
interface FastEthernet0/22
!
interface FastEthernet0/23
!
interface FastEthernet0/24
  switchport access vlan 101
!
interface GigabitEthernet0/1
!
interface GigabitEthernet0/2
!
interface Vlan1
  no ip address
  shutdown
!
interface Vlan30
  ip address 192.168.30.1 255.255.255.0
!
interface Vlan40
  ip address 192.168.40.1 255.255.255.0
!
interface Vlan101
  ip address 192.168.100.2 255.255.255.0
!
router ospf 2
  log-adjacency-changes
  network 192.168.100.0 0.0.0.255 area 0
  network 192.168.30.0 0.0.0.255 area 0
  network 192.168.40.0 0.0.0.255 area 0
!
ip classless
!
ip flow-export version 9
!
line con 0
```

```
!
line aux 0
!
line vty 0 4
 login
!
!
end
RS3560-B#
```

第六步：验证 PC 之间的连通性，如表 1-47 所示。

表 1-47　验证 PC 之间的连通性

PC	端口	PC	端口	结果	原因
PC1	A：1/1	PC2	A：1/9	通	
PC1	A：1/1	VLAN 100	A：1/24	通	
PC1	A：1/1	VLAN 101	B：0/0/24	通	
PC1	A：1/1	PC3	B：0/0/1	通	

七、注意事项和排错

在配置、使用 OSPF 协议时，可能会由于物理连接、配置错误等原因导致 OSPF 协议未能正常运行。因此，用户应注意以下要点。

1. 首先应该保证物理连接的正确无误。
2. 其次，保证接口和链路协议是 UP(使用 show interface 命令)。
3. 接下来，在各接口上配置不同网段的 IP 地址。
4. 最后，先启动 OSPF 协议(使用 router ospf 命令)再在相应接口配置所属 OSPF 域；接着，注意 OSPF 协议的自身特点——OSPF 骨干域(0 域)必须保证是连续的，如果不连续使用虚连接(virtual link)来保证，所有非 0 域只能通过 0 域与其他非 0 域相连，不允许非 0 域直接相连；边界三层交换机是指该三层交换机的一部分接口属于 0 域，而另外一部分接口属于非 0 域；对于广播网等多路访问网，需要选举指定的三层交换机作为 DR(指定路由器)。

八、思考题

本实验只体现了 area0 的配置方法，用户可以参考理论教材和用户手册尝试 OSPF 运行在多区域的配置方法。

OSPF 的配置命令如下：

```
default redistribute cost
default redistribute interval
default redistribute limit
default redistribute tag
default redistribute type
ip opsf authentication
```

```
ip ospf cost
ip opsf dead-interval
ip ospf enable area
ip ospf hello-interval
ip ospf passive-interface
ip ospf priority
ip ospf retransmit-interval
ip ospf transmit-delay
network
preference
redistribute ospfase
router id
router ospf
stub cost
virtuallink neighborid
show ip ospf
show ip ospfase
show ip ospf cumulative
show ip ospf database
show ip ospf interface
show ip ospf neighbor
show ip ospf routing
show ip ospf virtual-links
show ip protocols
debug ip ospf event
debug ip ospf lsa
debug ip ospf packet
debug ip ospf spf
```

实训项目七 多层交换机之间的动态 OSPF 配置

一、实训目的

1. 学会在多层交换机上配置动态 OSPF 路由，实现不同 VLAN 或子网之间的自动通信。
2. 理解动态路由和静态路由配置的区别。
3. 验证动态 OSPF 路由配置，诊断并解决路由配置中的问题。

二、实训设备

(1) DCS-3926S 交换机 2 台。
(2) DCRS-5526S 交换机 2 台。
(3) PC 机 4 台。
(4) 直通网线 7 根。

三、实训拓扑

该实验拓扑结构如图 1-45 所示。

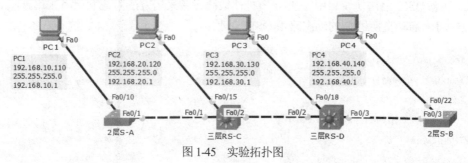

图 1-45 实验拓扑图

四、实训要求

参照如图 1-45 所示进行连线。
实训结果：PC1---ping---PC2----PC3----PC4   全通。

五、实训步骤

第一步：交换机 A、B、C、D 的配置如下。
A 划分 vl 10(9~12)、vl 20(13~16)，Trunk (1)。
B 划分 vl 30(17~20)、vl 40(21~24)，Trunk (3)。
C 划分 vl 10(9~12)、vl 20(13~16)、vl 100(2)，Trunk (1)。
D 划分 vl 30(17~20)、vl 40(21~24)、vl 101(2)，Trunk (3)。
第二步：设置三层交换机 C 启用三层路由功能。
设置 vl 10、vl 20、vl 100 的接口地址：

```
int vl 10    192.168.10.1/24;
int vl 20    192.168.20.1/24;
int vl 100  192.168.100.1/30。
```

第三步：设置三层交换机 D 启用三层路由功能。
设置 vl 30、vl 40、vl 101 的接口地址：

```
int vl 30    192.168.30.1/24;
int vl 40    192.168.40.1/24;
int vl 101  192.168.100.2/30。
```

第四步：根据连线位置正确配置 PC1、PC2、PC3、PC4 的地址(注意配置网关地址以及其与 VLAN 之间的对应关系)。
实训结果：PC1---ping---PC2----PC3----PC4   全通。
查看交换机状态：show vlan   show run   show ip route   show ip rip。

主要配置方法如下。

1. 交换机一般配置：略。

2. 动态 OSPF 路由协议启用方法如下(神州数码设备采用方法 1，思科设备采用方法 2)。

方法 1。启动交换机 C 的动态路由 OSPF 协议：

```
C(config)#router ospf
C(config-Router-ospf)#exit
C(config)#int vl 10
C(config-if-vlan10)#ip ospf enable area 0
C(config)#int vl 20
C(config-if-vlan20)#ip ospf enable area 0
C(config)#int vl 100
C(config-if-vlan100)#ip ospf enable area 0
```

验证：sh ip rip   sh ip route。

启动交换机 D 的动态路由 OSPF 协议：

```
D(config)#router ospf
D(config-Router-ospf)#exit
D(config)#int vl 30
D(config-if-vlan30)#ip ospf enable area 0
D(config)#int vl 40
D(config-if-vlan40)#ip ospf enable area 0
D(config)#int vl 101
D(config-if-vlan101)#ip ospf enable area 0
```

验证：sh ip rip   sh ip route。

方法 2。启动交换机 C 的动态路由 OSPF 协议：

```
C(config)#router ospf
C(config-Router)#network 192.168.10.0 255.255.255.0 area 0
C(config-Router)#network 192.168.20.0 255.255.255.0 area 0
C(config-Router)#network 192.168.100.0 255.255.255.0 area 0
```

启动交换机 D 的动态路由 OSPF 协议：

```
C(config)#router ospf
C(config-Router)#network 192.168.30.0 255.255.255.0 area 0
C(config-Router)#network 192.168.40.0 255.255.255.0 area 0
C(config-Router)#network 192.168.100.0 255.255.255.0 area 0
```

验证：sh ip rip   sh ip route。

第 2 章

路由器实验

实验一 路由器接口简介

一、实验目的

1. 熟悉路由器各接口的外观。
2. 理解接口的功能。
3. 掌握接口的表示方法。

二、相关知识

路由器是三层设备,其核心功能在于进行精确的路径选择以及实现广域网的连接。与交换机相比,路由器的接口数量虽然较少,但其功能却更为强大和多样化。这些多样化的功能在外观上体现为接口和模块类型的丰富性。当然,不同功能和模块的路由器在价格上存在显著差异。高端路由器通常采用模块化设计,支持多种类型的模块,从而满足更为复杂和多样化的网络需求。

三、实验设备

DCR 1702 1 台。

四、实验拓扑

该实验拓扑结构如图 2-1 所示。

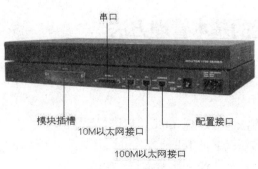

图 2-1 实验拓扑图

五、实验要求

1. 以太网接口有 100Mbps 和 10Mbps 之分。
2. 注意观察接口和模块上的标志。
3. 注意路由器插槽上标号,从靠近电源开始依次为 0、1、2、3。
4. 路由器各接口从 0 开始。

六、实验步骤

第一步:连接一个固定的 Console 接口,用于本地配置路由器或进行路由器的软件升级。
第二步:使用 10/100Mbps 自适应的 RJ45 以太网接口,这些接口主要用于连接以太网。
第三步:根据需要,选择模块扩展插槽进行扩展。DCR-1751 配备一个模块扩展插槽,而 DCR-1750 则有三个模块扩展插槽,具体模块的说明如表 2-1 所示。

表 2-1 模块说明

序号	模块名称	模块描述	Slot 1	Slot 2	Slot 3
1	MR-WIC-1ETH	单路 10Mbps 以太网接口卡	Yes	Yes	/
2	MR-WIC-2ETH	双路 10Mbps 以太网接口卡	Yes	Yes	/
3	MR-WIC-1T	单路同步/异步串口卡	Yes	Yes	/
4	MR-WIC-2T	双路同步/异步串口卡	Yes	Yes	/
5	MR-WIC-1E1T	单路 10Mbps 以太网+单同步/异步串口卡	Yes	Yes	/
6	MR-WIC-1CE1	单路 E1 接口卡	/	Yes	/
7	MR-WIC-1B-S/T	单路 ISDN BRI S/T 接口卡	/	Yes	/
8	MR-VIC-2FXS	双口 FXS 语音接口卡	Yes	Yes	Yes
9	MR-VIC-2FXO	双口 FXO 语音接口卡	Yes	Yes	Yes
10	MR-VIC-2E&M	双口 E&M 语音接口卡	Yes	Yes	Yes
11	MR-EIC-8A	8 路异步通信卡	Yes	Yes	Yes

七、注意事项和排错

1. 加装和拆卸模块一定要先关闭电源。
2. 串口线不要带电插拔。

实验二 路由器的基本管理方法

一、实验目的

1. 掌握带外的管理方法:通过 Console 接口进行配置。
2. 掌握带内的管理方法:通过 Telnet 方式进行配置。
3. 掌握带内的管理方法:通过 Web 界面进行配置。

二、相关知识

路由器管理包括通过 CLI 或 GUI 访问设备，进行初始配置(如设置密码和 IP 地址)、执行日常任务(包括接口配置、路由协议设置、ACL 和 QoS 配置、NAT 和 VPN 管理)，以及监控和维护设备状态。此外，还需采取安全措施(如更改默认密码和使用 SSH)，并进行故障排除，以保障网络稳定和安全。设备的初始配置通常通过 Console 接口完成。远程管理则通常通过带内方式进行，带内管理需要在相应接口配置 IP 地址并开启相关服务后才能进行。

三、实验设备

1. DCR-1702 1 台。
2. DCR-2611 1 台。
3. PC 机 1 台。
4. Console 线缆、网线各 1 条。

四、实验拓扑

该实验拓扑结构如图 2-2 所示。

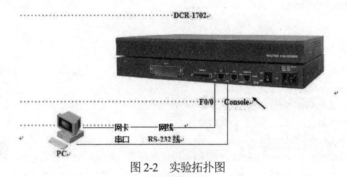

图 2-2　实验拓扑图

五、实验要求

设备接口配置如表 2-2 所示。

表 2-2　设备接口配置

DCR-1702		PC	
Console		串口	
F0/0	192.168.2.1	网卡	192.168.2.2

六、实验步骤

带外管理方法(本地管理)。

第一步：将配置线的一端与路由器的 Console 口相连，另一端与 PC 的串口相连，如图 2-2 所示。

第二步：在 PC 上运行终端仿真程序。选择"开始"|"程序"|"附件"|"通讯"|"超级终端"命令，打开如图 2-3 所示的"属性"对话框，设置终端的硬件参数。

图 2-3 "属性"对话框

第三步：路由器加电，超级终端会显示路由器自检信息，自检结束后显示命令提示"Press RETURN to get started"。

```
System Bootstrap, Version 0.1.8
Serial num:8IRT01V11B01000054 ,ID num:000847
Copyright (c) 1996-2000 by China Digitalchina CO.LTD
DCR-1700 Processor MPC860T @ 50Mhz
The current time: 2067-9-12 6:31:30
Loading DCR-1702.bin……
Start Decompress DCR-1702.bin
################################################################################
################################################################################
################################################################################
########################
Decompress 3587414 byte,Please wait system up.
Digitalchina Internetwork Operating System Software
DCR-1700 Series Software , Version 1.3.2E, RELEASE SOFTWARE
System start up OK
Router console 0 is now available
Press RETURN to get started
```

第四步：按回车键进入用户配置模式。DCR-1702 系列路由器出厂时未设置密码，用户按回车键后将直接进入普通用户模式，在该模式下可以使用权限允许范围内的命令。如需帮助，可以输入"？"。如果要进入超级用户模式，输入 enable 并按回车键即可。这时用户将获得最高权限，可以对路由器进行全面配置。

```
Router-A>enable                    ! 进入特权模式
    Router-A#?                     ! 查看可用的命令
```

Cd		-- Change directory
chinese		-- Help message in Chinese
chmem		-- Change memory of system
chram		-- Change memory
clear		-- Clear something
config		-- Enter configurative mode
connect		-- Open a outgoing connection
copy		-- Copy configuration or image data
date		-- Set system date
debug		-- Debugging functions
delete		-- Delete a file
dir		-- List files in flash memory
disconnect		-- Disconnect an existing outgoing network connection
download		-- Download with ZMODEM
enable		-- Turn on privileged commands
english		-- Help message in English
enter		-- Turn on privileged commands
exec-script		-- Execute a script on a port or line
exit		-- Exit / quit
format		-- Format file system
help		-- Description of the interactive help system
history		-- Look up history
Router-A#ch?		！使用"？"获取帮助
chinese		-- Help message in Chinese
chmem		-- Change memory of system
chram		-- Change memory
Router-A#chinese		！设置中文帮助
Router-A#?		！再次查看可用命令
cd		-- 改变当前目录
chinese		-- 中文帮助信息
chmem		-- 修改系统内存数据
chram		-- 修改内存数据
clear		-- 清除
config		-- 进入配置态
connect		-- 打开一个向外的连接
copy		-- 拷贝配置方案或内存映像
date		-- 设置系统时间
debug		-- 分析功能
delete		-- 删除一个文件
dir		-- 显示闪存中的文件
disconnect		-- 断开活跃的网络连接
download		-- 通过 ZMODEM 协议下载文件
enable		-- 进入特权方式
english		-- 英文帮助信息

enter	-- 进入特权方式
exec-script	-- 在指定端口运行指定的脚本
exit	-- 退回或退出
format	-- 格式化文件系统
help	-- 交互式帮助系统描述
history	-- 查看历史
keepalive	-- 保活探测
--More--	

带内远程的管理方法(Telnet 方式)。

第五步：设置路由器以太网接口地址并验证。

```
Router>en
Router>enable
Router#config t
Router#config terminal
Router(config)#interface f0/0
Router(config-if)#ip address 192.168.2.1 255.255.255.0
Router(config-if)#no shutdown
Router(config-if)#^Z
Router#
Router#show interfaces f0/0
FastEthernet0/0 is up, line protocol is down (disabled)
Hardware is Lance, address is 0000.0c72.1201 (bia 0000.0c72.1201)
Internet address is 192.168.2.1/24
MTU 1500 bytes, BW 100000 Kbit, DLY 100 usec,
reliability 255/255, txload 1/255, rxload 1/255
Encapsulation ARPA, loopback not set
ARP type: ARPA, ARP Timeout 04:00:00,
Last input 00:00:08, output 00:00:05, output hang never
Last clearing of "show interface" counters never
Input queue: 0/75/0 (size/max/drops); Total output drops: 0
Queueing strategy: fifo
Output queue :0/40 (size/max)
5 minute input rate 0 bits/sec, 0 packets/sec
5 minute output rate 0 bits/sec, 0 packets/sec
0 packets input, 0 bytes, 0 no buffer
Received 0 broadcasts, 0 runts, 0 giants, 0 throttles
0 input errors, 0 CRC, 0 frame, 0 overrun, 0 ignored, 0 abort
0 input packets with dribble condition detected
0 packets output, 0 bytes, 0 underruns
0 output errors, 0 collisions, 1 interface resets
0 babbles, 0 late collision, 0 deferred
0 lost carrier, 0 no carrier
0 output buffer failures, 0 output buffers swapped out
```

第六步：设置 PC 的 IP 地址(如图 2-4 所示)并测试连通性。

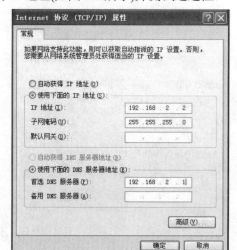

图 2-4　设置 IP 地址

使用 ping 命令测试连通性，如图 2-5 所示。

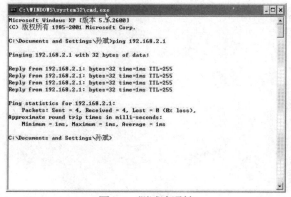

图 2-5　测试连通性

第七步：在 PC 上 Telnet 到路由器。运行 telnet 192.168.2.1，如图 2-6 所示。

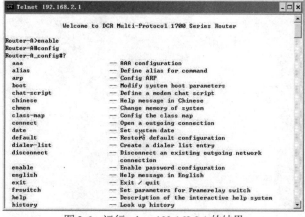

图 2-6　运行 telnet 192.168.2.1 的结果

带内远程的管理方法：(Web 方式)。

第八步：由于 DCR-1702 不支持 Web 管理方式，本例将以 DCR-2611 为例进行配置。按照相同的配置方法，将以太网接口地址配置为 192.168.2.2/24。

```
Router#config terminal
Router(config)#interface fastEthernet 0/0
Router(config-if)#ip address 192.168.2.2 255.255.255.0
Router(config-if)#no shutdown
Router(config-if)#exit
Router(config)#ip http server
Router(config)#^Z
Router#
Router#show running-config
Building configuration...
Current configuration : 555 bytes
!
version 12.4
service timestamps debug datetime msec
service timestamps log datetime msec
no service password-encryption
!
hostname Router
!
boot-start-marker
boot-end-marker
!
!
no aaa new-model
memory-size iomem 5
!
!
ip cef
no ip domain lookup
!
!
interface FastEthernet0/0
 ip address 192.168.2.2 255.255.255.0
 duplex auto
 speed auto
!
ip http server
no ip http secure-server
!
control-plane
```

```
!
line con 0
  exec-timeout 0 0
  logging synchronous
line aux 0
line vty 0 4
!
!
end
Router#
```

七、注意事项和排错

1. 在超级终端中的配置是对路由器进行操作，此时的 PC 仅作为输入输出设备使用。
2. 在使用 Telnet 和 Web 方式管理路由器时，应先测试连通性。

八、思考题

1. 带内和带外管理方式各有什么优点和缺点？
2. Telnet 和 Web 的端口号是什么？

九、课后练习

将所有设备的 IP 地址更改为 10.0.0.0/24 网段内的地址，然后重复本实验的配置过程。

实验三　路由器的基本配置

一、实验目的

1. 通过实验配置，理解路由器的工作原理。
2. 掌握路由器的基本配置。
3. 诊断并排除配置路由器过程中出现的网络问题。

二、相关知识

路由器的基本配置包括接口 IP 地址分配、路由协议配置以及安全设置等。这些配置涉及理解路由器的工作原理、掌握配置命令和策略，以确保网络的连通性、性能和安全。在执行实际配置之前，本节将学习一些基本配置方法，主要包括配置路由器的名称、接口地址以及特权模式密码等内容。

三、实验设备

1. DCR-1751 2 台。
2. CR-V35MT 1 条。
3. CR-V35FC 1 条。
4. 网线 2 根。

四、实验拓扑

该实验拓扑结构如图 2-7 所示。

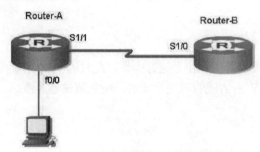

图 2-7　实验拓扑图

五、实验要求

在实验中，路由器(Router-A 和 Router-B)接口的配置如表 2-3 所示。

表 2-3　路由器各接口配置表

Router-A			Router-B		
接口	类型	IP 地址	接口	类型	IP 地址
S1/1	DCE	192.168.2.1	S0/0	DTE	192.168.2.2
F0/0		192.168.2.1			

六、实验步骤

Router-A 的基本配置步骤如下。

第一步：恢复出厂设置。

```
hgbnnnnnnnnnnnnnnnnnnnRouter>enable
Router#show running-config
Building configuration...
Current configuration:
!
!version 1.3.2E
<省略....>
Router#erase startup-config
Erasing the nvram filesystem will remove all configuration files! Continue? [confirm]
```

[OK]
Erase of nvram: complete

Router#reload
System configuration has been modified. Save? [yes/no]: y
Building configuration...
[OK]

第二步：设置接口 IP 地址、DCE 的时钟频率以及验证。

Router>enable
Router#config terminal
Router(config)#hostname Router-A
Router-A(config)#
Router-A((config)#int s1/1
Router-A((config-if)#ip address 192.168.2.1 255.255.255.0
Router-A((config-if)#clock rate 64000
Router-A((config-if)#no shutdown
Router-A((config-if)#^Z
Router-A(#
Router-A(#show interfaces s1/1
Serial1/1 is administratively down, line protocol is down
　　Hardware is M4T
　　Internet address is 192.168.2.1/24
　　MTU 1500 bytes, BW 1544 Kbit, DLY 20000 usec,
　　　　reliability 255/255, txload 1/255, rxload 1/255
　　Encapsulation HDLC, crc 16, loopback not set
　　Keepalive set (10 sec)
　　Restart-Delay is 0 secs
　　Last input never, output never, output hang never
　　Last clearing of "show interface" counters never
　　Input queue: 0/75/0/0 (size/max/drops/flushes); Total output drops: 0
　　Queueing strategy: weighted fair
　　Output queue: 0/1000/64/0 (size/max total/threshold/drops)
　　　　Conversations　　0/0/256 (active/max active/max total)
　　　　Reserved Conversations 0/0 (allocated/max allocated)
　　　　Available Bandwidth 1158 kilobits/sec
　　5 minute input rate 0 bits/sec, 0 packets/sec
　　5 minute output rate 0 bits/sec, 0 packets/sec
　　　　0 packets input, 0 bytes, 0 no buffer
　　　　Received 0 broadcasts, 0 runts, 0 giants, 0 throttles
　　　　0 input errors, 0 CRC, 0 frame, 0 overrun, 0 ignored, 0 abort
　　　　0 packets output, 0 bytes, 0 underruns
　　　　0 output errors, 0 collisions, 1 interface resets
　　　　0 output buffer failures, 0 output buffers swapped out
　　　　1 carrier transitions　　　DCD=up　DSR=up　DTR=up　RTS=up　CTS=up

```
Router-A#configure terminal
Router-A(config)#interface f0/0
Router-A(config-if)#ip address 192.168.2.1 255.255.255.0
Router-A(config-if)#no shutdown
Router-A(config-if)#^Z
Router-A#show interface f0/0
FastEthernet0/0 is up, line protocol is up
   Hardware is AmdFE, address is cc02.32dc.0000 (bia cc02.32dc.0000)
   Internet address is 192.168.2.1/24
   MTU 1500 bytes, BW 100000 Kbit, DLY 100 usec,
      reliability 255/255, txload 1/255, rxload 1/255
   Encapsulation ARPA, loopback not set
   Keepalive set (10 sec)
   Full-duplex, 100Mb/s, 100BaseTX/FX
   ARP type: ARPA, ARP Timeout 04:00:00
   Last input never, output never, output hang never
   Last clearing of "show interface" counters never
   Input queue: 0/75/0/0 (size/max/drops/flushes); Total output drops: 0
   Queueing strategy: fifo
   Output queue: 0/40 (size/max)
   5 minute input rate 0 bits/sec, 0 packets/sec
   5 minute output rate 0 bits/sec, 0 packets/sec
      0 packets input, 0 bytes
      Received 0 broadcasts, 0 runts, 0 giants, 0 throttles
      0 input errors, 0 CRC, 0 frame, 0 overrun, 0 ignored
      0 watchdog
      0 input packets with dribble condition detected
      60 packets output, 6457 bytes, 0 underruns
      0 output errors, 0 collisions, 0 interface resets
```

第三步：设置特权模式密码。

```
Router-A(config)#enable password digitalchina
Router-A(config)#^Z
Router-A#exit
Router-A>enable
Password:
Access deny !
Router-A>enable
Password:
Router-A#
```

第四步：保存。

```
Router-A#write
Saving current configuration...
OK!
```

第五步：查看配置序列。

```
Router-A#show running-config
Building configuration...

Current configuration : 936 bytes
!
version 12.4
service timestamps debug datetime msec
service timestamps log datetime msec
no service password-encryption
!
hostname Router-A
!
boot-start-marker
boot-end-marker
!
enable password digitalchina
!
no aaa new-model
memory-size iomem 5
!
ip cef
no ip domain lookup
!
interface FastEthernet0/0
  ip address 192.168.2.1 255.255.255.0
  duplex auto
  speed auto
!
interface Serial1/0
  no ip address
  shutdown
  serial restart-delay 0
  clock rate 64000
!
interface Serial1/1
  ip address 192.168.2.1 255.255.255.0
  shutdown
  serial restart-delay 0
  clock rate 64000
!
interface Serial1/2
  no ip address
  shutdown
  serial restart-delay 0
```

```
!
interface Serial1/3
 no ip address
 shutdown
 serial restart-delay 0
!
ip http server
no ip http secure-server
!
control-plane
!
line con 0
 exec-timeout 0 0
 logging synchronous
line aux 0
line vty 0 4
 login
!
end
```

Router-B 的配置步骤如下(命令解释参照 Router-A 的配置)。

第一步：恢复出厂设置。

```
Router>enable
Router#show running-config
Building configuration...
Current configuration:
!
!version 1.3.2E
<省略....>
Router#erase startup-config
Erasing the nvram filesystem will remove all configuration files! Continue? [confirm]
[OK]
Erase of nvram: complete
Router#reload
System configuration has been modified. Save? [yes/no]: y
Building configuration...
[OK]
```

第二步：设置 IP 地址并进行验证。

```
Router>enable
Router#config terminal
Router(config)#hostname Router-B
Router-B(config)#interface s1/1
Router-B(config-if)#ip address 192.168.2.2 255.255.255.0
```

```
Router-B(config-if)#no shutdown
Router-B(config-if)#^Z
Router-B#show interfaces s1/1
Serial1/0 is up, line protocol is up
  Hardware is M4T
  Internet address is 192.168.2.2/24
  MTU 1500 bytes, BW 1544 Kbit, DLY 20000 usec,
     reliability 255/255, txload 1/255, rxload 1/255
  Encapsulation HDLC, crc 16, loopback not set
  Keepalive set (10 sec)
  Restart-Delay is 0 secs
  Last input never, output never, output hang never
  Last clearing of "show interface" counters never
  Input queue: 0/75/0/0 (size/max/drops/flushes); Total output drops: 0
  Queueing strategy: weighted fair
  Output queue: 2/1000/64/0 (size/max total/threshold/drops)
     Conversations   1/1/256 (active/max active/max total)
     Reserved Conversations 0/0 (allocated/max allocated)
     Available Bandwidth 1158 kilobits/sec
  5 minute input rate 0 bits/sec, 0 packets/sec
  5 minute output rate 0 bits/sec, 0 packets/sec
     0 packets input, 0 bytes, 0 no buffer
     Received 0 broadcasts, 0 runts, 0 giants, 0 throttles
     0 input errors, 0 CRC, 0 frame, 0 overrun, 0 ignored, 0 abort
     2 packets output, 341 bytes, 0 underruns
     0 output errors, 0 collisions, 1 interface resets
     0 output buffer failures, 0 output buffers swapped out
     2 carrier transitions     DCD=down  DSR=down  DTR=up  RTS=up  CTS=down
```

第三步：保存。

```
Router-B#write
Saving current configuration...
OK!
```

第四步：查看配置序列。

```
Router-B#show running-config
Building configuration...
Current configuration : 844 bytes
!
version 12.4
service timestamps debug datetime msec
service timestamps log datetime msec
no service password-encryption
!
hostname Router-B
```

```
!
boot-start-marker
boot-end-marker
!
!
no aaa new-model
memory-size iomem 5
!
ip cef
no ip domain lookup
!
interface FastEthernet0/0
 no ip address
 shutdown
 duplex auto
 speed auto
!
interface Serial1/0
 no ip address
 shutdown
 serial restart-delay 0
!
interface Serial1/1
 ip address 192.168.2.2 255.255.255.0
 serial restart-delay 0
!
interface Serial1/2
 no ip address
 shutdown
 serial restart-delay 0
!
interface Serial1/3
 no ip address
 shutdown
 serial restart-delay 0
!
no ip http server
no ip http secure-server
!
control-plane
!
line con 0
 exec-timeout 0 0
 logging synchronous
line aux 0
```

```
line vty 0 4
!
end
```

第五步：测试连通性。

```
Router-B#ping 192.168.2.1
Type escape sequence to abort.
Sending 5, 100-byte ICMP Echos to 192.168.2.1, timeout is 2 seconds:
!!!!!
Success rate is 100 percent (5/5), round-trip min/avg/max = 20/28/44 ms
```

七、注意事项和排错

1. CR-V35FC 所连的接口为 DCE，需要配置时钟频率，CR-V35MT 所连的接口为 DTE。

2. 查看接口状态，如果接口状态显示为"DOWN"，通常可能是线缆故障；如果协议状态显示为"DOWN"，可能是时钟频率未配置正确，或者是两端封装协议不一致。

八、配置序列

已在步骤中列出。

九、思考题

1. 如果要将特权模式密码以密文显示，应使用什么参数(可以通过输入"？"进行查看)？
2. 如果要表示插槽 2 上的第 2 个快速以太网接口，应该怎么写？

十、课后练习

将所有地址更改为 10.0.0.0/24 这个网段，并重复上述配置。

实验四　路由器的文件维护

一、实验目的

1. 掌握路由器的软件升级方法。
2. 备份和还原路由器的配置文件。

二、相关知识

路由器的文件维护实验是网络设备管理中的一项重要内容，涉及路由器操作系统的备份、恢复、升级以及配置文件的维护等。正常情况下，可以通过 TFTP 或 FTP 方式进行操作，当设备无法正常启动时，则可以通过 ZMODEN 方式进行恢复。

三、实验设备

1. DCR-1751 1 台。
2. PC 1 台。
3. TFTP 和 FTP 软件。

四、实验拓扑

该实验拓扑结构如图 2-8 所示。

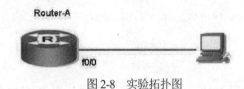

图 2-8　实验拓扑图

五、实验要求

在实验中，各设备接口的配置如表 2-4 所示。

表 2-4　设备接口配置

Router-A		PC	
接口	IP 地址	网卡	IP 地址
F0/0	192.168.2.1		192.168.2.10

六、实验步骤

1. TFTP 方式(采用 UDP 协议，适合本地操作)

第一步：设置 PC 的网卡地址为 192.168.2.10，并安装 3Cdaeom 软件，设置 TFTP Server 配置，如图 2-9 所示。

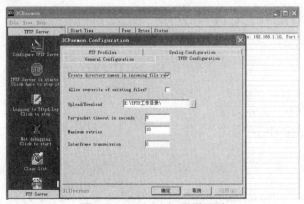

图 2-9　TFTP Server 设置界面

第二步：设置 DCR-1751 的 F0/0 接口地址为 192.168.2.1，并测试连通性(参照本章实验)。

```
Router-A#config terminal
Router-A(config)#interface f0/0
Router-A(config-if)#ip address 192.168.2.1 255.255.255.0
Router-A(config-if)#no shutdown
Router-A(config-if)#^Z
Router-A#show interface f0/0
FastEthernet0/0 is up, line protocol is up
    Hardware is AmdFE, address is cc02.32dc.0000 (bia cc02.32dc.0000)
    Internet address is 192.168.2.1/24
    MTU 1500 bytes, BW 100000 Kbit, DLY 100 usec,
        reliability 255/255, txload 1/255, rxload 1/255
    Encapsulation ARPA, loopback not set
    Keepalive set (10 sec)
<省略….>
Router-A#ping 192.168.2.10
PING 192.168.2.10 (192.168.2.1): 56 data bytes
!!!!!
--- 192.168.2.1 ping statistics ---
5 packets transmitted, 5 packets received, 0% packet loss
round-trip min/avg/max = 20/22/30 ms
```

第三步：查看路由器文件，并将配置文件下载到 TFTP 服务器上。

```
Router-A#dir nvram:
Directory of nvram:/
    124  -rw-       936            <no date>   startup-config
    125  ----        24            <no date>   private-config
      1  -rw-         0            <no date>   ifIndex-table
129016 bytes total (126980 bytes free)
Router-A#copy nvram:startup-config tftp:
Remote-server ip address[]?192.168.2.10
Destination file name[startup-config]?
#
TFTP:successfully send 2 blocks ,516 bytes
```

第四步：使用写字板打开下载后的配置文件，修改机器名称，然后将其上传到路由器中。重新启动后，使用 show 命令可以观察到机器名称已经被修改。

2. FTP 方式(采用 TCP 协议，适合远程操作)

路由器作为 FTP 客户端的配置方法与 TFTP 类似，只需将命令中的 TFTP 替换为 FTP 即可。接下来，我们将学习如何将路由器配置为服务器端以更新文件。

第一步：配置接口地址，参见 TFTP 方式的第二步。

第二步：开启 FTP 服务器功能，并配置 FTP 服务器用户名和密码。

```
Router-A(Config)#ip ftp server
Router-A(Config)#ip ftp username router
Router-A(Config)#ip ftp password 0 digitalchina
```

第三步：在 PC 上打开 IE 浏览器作为客户端进行上传或下载。

在浏览器地址栏输入网址：ftp://router:digitalchina@192.168.2.1。

3. 启动到 monitor

当路由器的软件被破坏无法启动的时候，可以在启动过程中按 Ctrl + Break 组合键，进入 MONITOR 模式，使用 ZMODEM 方式恢复文件，所谓 ZMODEM 方式指的是从路由器的 Console 端口以波特率规定的速率通过 PC 的串口传输文件的一种方式，不需要网线。

第一步：重启路由器，在启动过程中按 Ctrl + Break 组合键，进入 MONITOR 模式。

```
System Bootstrap, Version 0.1.8
Serial num:8IRT01V11B01000054 ,ID num:000847
Copyright (c) 1996-2000 by China Digitalchina CO.LTD
DCR-1700 Processor MPC860T @ 50Mhz
The current time: 2067-9-12 4:44:13
              Welcome to DCR Multi-Protool 1700 Series Router
monitor#
```

第二步：设置传输方式。

```
monitor#download c0 routerb            ！设置从 Console 传输文件
Speed: [9600]                          ！选择波特率
```

第三步：执行以下操作。

(1) 启动 Windows 系统，选择"开始"|"程序"|"附件"|"通讯"|"超级终端"命令。

(2) 在打开的"超级终端"窗口中选择"传送"|"发送文件"命令，打开"发送文件"对话框，如图 2-10 所示。

图 2-10 "发送文件"对话框

(3) 单击"发送文件"对话框中的"浏览"按钮，在打开的对话框中手动选择备份的路由器软件，然后单击"发送"按钮，打开如图 2-11 所示的界面。

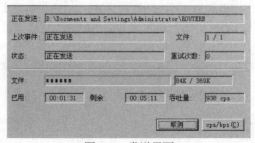

图 2-11 发送界面

4. 恢复遗失的密码

当密码遗忘时，可以进入 MONITOR 模式并执行以下操作清除密码：

monitor#no passwd

七、注意事项和排错

1. 路由器和 PC 直接相连时使用交叉线。
2. 关闭 PC 上的防火墙。
3. 在实际工作中，通常使用日期或功能等标明配置文件。
4. 使用 ZMODEM 方式恢复文件，当文件比较大时比较耗时。

八、配置序列

无。

九、思考题

1. TFTP 和 FTP 这两种方式有什么区别？
2. 在 MONITOR 模式下，为什么不能使用 TFTP 方式？

十、课后练习

写出使用 FTP 上传配置文件的过程。

实验五 单臂路由实验

一、实验目的

1. 了解路由器各接口的外观。
2. 掌握接口的功能。
3. 掌握接口的表示方法。

二、相关知识

单臂路由(Single Arm Router)是一种网络技术通过在路由器的一个物理接口上运行多个逻辑(虚拟)接口，实现不同 VLAN 之间的路由。配置单臂路由的步骤是：①在交换机上创建 VLAN。②在路由器上创建子接口。③为每个接口分配唯一的 VLAN 和 IP 地址，并在子接口上配置 802.1q 封装。通过这些配置，不同 VLAN 之间的通信得以实现。

单臂路由实验主要应用于以下场景。

(1) 简化网络结构。在只有少量 VLAN 需要互联的小型网络中，单臂路由可以简化网络结构，减少对物理设备和布线的需求。

(2) 办公网络隔离。在大型办公环境中，单臂路由可实现不同部门之间的网络隔离，同时确保必要的通信。

(3) 数据中心。在数据中心，单臂路由可用于实现服务器与存储网络之间的隔离和互联。

三、实验设备

1. 路由器 1 台。
2. 二层交换机 3 台。
3. PC 机 5 台。
4. 直通网线若干。

四、实验拓扑

该实验拓扑结构如图 2-12 所示。

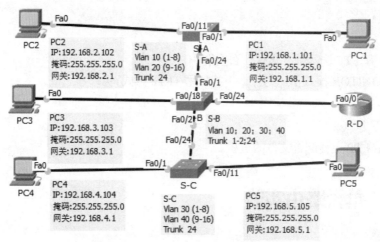

图 2-12　单臂路由实验拓扑图

五、实验要求

1. 设备配置

按图 2-12 所示进行连线配置。

地址		gateway	Mask	连线
PC1	192.168.1.101	192.168.1.1	255.255.255.0	A-0/1
PC2	192.168.2.102	192.168.2.1	255.255.255.0	A-0/11
PC3	192.168.3.103	192.168.3.1	255.255.255.0	B-0/18
PC4	192.168.4.104	192.168.4.1	255.255.255.0	C-0/1
PC5	192.168.5.105	192.168.5.1	255.255.255.0	C-0/11

2. 交换机设置

在交换机 A 上划分两个 VLAN，端口为：

VLAN	端口成员
10	1~8
20	9~16
trunk 口	0/24

在交换机 C 上划分两个 VLAN，端口为：

VLAN	端口成员
30	1~8
40	9~16
trunk 口	0/24

在交换机 B 上：
VLAN 10、VLAN 20、VLAN 30 和 VLAN 40 不设置端口成员。

trunk 口	0/1;0/2;0/24

3. 交换机连接

交换机 S-B 的 0/24 口连接路由器 R-D 的 F 0/0 口。
交换机 S-A 的 0/24 口连接交换机 S-B 的 0/1 口。
交换机 S-C 的 0/24 口连接交换机 S-B 的 0/2 口。

4. 路由器配置

分别在 F0/0.10、F0/0.20、F0/0.30、F0/0.40、F0/0.1 子接口配置 IP，作为网关，并进行 dot1Q 封装 VLAN。

确保 PC1、PC2、PC3、PC4 和 PC5 之间通信畅通。
思考：PC6　连线 S-B　0/6　IP 如何设置可以通。

六、实验步骤

交换机配置：

```
S-A(config)#vlan 10
S-A(config-vlan)#vlan 20
S-A(config-vlan)#ex
S-A(config)#interface range fastEthernet 0/1-8
S-A(config-if-range)#switchport access vlan 10
S-A(config-vlan)#ex
S-A(config)#interface range fastEthernet 0/9-16
S-A(config-if-range)#switchport access vlan 20
```

S-A(config-if-range)#ex

S-A(config)#interface fastEthernet   0/24

S-A(config-if)#switchport mode trunk

S-B(config)#vlan 10

S-B(config-vlan)#vlan 20

S-B(config-vlan)#vlan 30

S-B(config-vlan)#vlan 40

S-B(config-vlan)#ex

S-B(config)#interface fastEthernet 0/1

S-B(config-if)#switchport mode trunk

S-B(config)#interface fastEthernet 0/2

S-B(config-if)#switchport mode trunk

S-B(config)#interface fastEthernet 0/24

S-B(config-if)#switchport mode trunk

S-C(config)#vl

S-C(config)#vlan 30

S-C(config-vlan)#vlan 40

S-C(config-vlan)#ex

S-C(config)#interface range fastEthernet 0/1-8

S-C(config-if-range)#switchport access vlan 30

S-C(config)#interface range fastEthernet 0/9-16

S-C(config-if-range)#switchport access vlan 40

S-C(config-if-range)#ex

S-C(config)#interface fastEthernet 0/24

S-C(config-if)#switchport mode trunk

路由器配置：
直接进入 F0/0.10 和 F0/0.20 子接口配置 IP 并进行 dot1Q 封装。

R-D(config)#interface fastEthernet 0/0.1

R-D(config-subif)#encapsulation dot1Q ?

   <1-1005>   IEEE 802.1Q VLAN ID

R-D(config-subif)#encapsulation dot1Q 1

R-D(config-subif)#ip address 192.168.3.1 255.255.255.0

R-D(config-subif)#interface fastEthernet 0/0.10

R-D(config-subif)#encapsulation dot1Q 10

R-D(config-subif)#ip address 192.168.1.1 255.255.255.0

R-D(config-subif)#interface fastEthernet 0/0.20

R-D(config-subif)#encapsulation dot1Q 20

R-D(config-subif)#ip address 192.168.2.1 255.255.255.0

R-D(config-subif)#interface fastEthernet 0/0.30
R-D(config-subif)#encapsulation dot1Q 30
R-D(config-subif)#ip address 192.168.4.1 255.255.255.0
R-D(config-subif)#interface fastEthernet 0/0.40
R-D(config-subif)#encapsulation dot1Q 40
R-D(config-subif)#ip address 192.168.5.1 255.255.255.0
R-D#show IP route 验证路由表

实验：验证 PC1------ping-------PC2，PC3，PC4，PC5　通。

七、课后练习

场景描述：假设你是一家小型企业的网络管理员，公司有两个部门(市场营销部和财务部)，分别位于不同的 VLAN 中。为了实现两个部分之间的通信，你需要配置一个单臂路由器。

1. 网络设计：确定 VLAN 分配和 IP 地址规划，并完成网络的连接设计。
2. VLAN 配置：在交换机上创建两个 VLAN，并将相应的端口分配到对应的 VLAN 中。
3. 单臂路由配置：在路由器上创建子接口，为每个子接口分配 IP 地址，并进行 802.1q 封装。
4. PC 配置：为连接到各个 VLAN 的 PC 配置 IP 地址、网关等网络信息。
5. 验证：从一个 VLAN 中的 PC 尝试 ping 另一个 VLAN 中的 PC，以确认网络连通性。

实验六　路由器静态路由配置

一、实验目的

1. 学习静态路由的定义、工作原理以及配置方法，理解静态路由与动态路由的区别。
2. 通过实际操作，掌握如何在路由器上配置静态路由，包括指定目标网络、子网掩码和下一跳地址。
3. 学习如何测试和验证网络的连通性，确保配置的静态路由能够正确转发数据包。

二、相关知识

静态路由是由网络管理员手动配置的路由信息，用于指示路由器如何到达特定的网络目的地。在小规模环境中，静态路由是一种理想的选择。静态路由开销较低，但灵活性不足，适用于相对稳定的网络。

路由器静态路由配置主要应用于以下场景。

(1) 小型网络：在小型网络中，由于网络结构简单，使用静态路由可以轻松实现网络互连。

(2) 特定路径控制：当网络管理员需要精确控制数据包的传输路径时，静态路由是一个非常有效的工具。

(3) 默认路由应用：在网络出口处，通常会配置默认路由，将所有对外网络的请求转发至互联网网关。

三、实验设备

1. DCR-1751 3 台。
2. CR-V35FC 1 条。
3. CR-V35MT 1 条。

四、实验拓扑

该实验拓扑结构如图 2-13 所示。

图 2-13　实验拓扑图

五、实验要求

实验中各设备接口的配置如表 2-5 所示。

表 2-5　设备接口配置

Router-A		Router-B		Router-C	
S1/1(DCE)	192.168.5.1	S1/0(DTE)	192.168.5.2	F0/0	192.168.2.2
F0/0	192.168.0.1	F0/0	192.168.2.1	E1/0	192.168.3.1

六、实验步骤

第一步：按照表 2-5 配置所有接口的 IP 地址，确保所有接口均处于 up 状态，并测试网络连通性。

第二步：查看 Router-A 的路由表。

```
Router-A#sh ip route
Codes: C - connected, S - static, R - RIP, M - mobile, B - BGP
       D - EIGRP, EX - EIGRP external, O - OSPF, IA - OSPF inter area
       N1 - OSPF NSSA external type 1, N2 - OSPF NSSA external type 2
       E1 - OSPF external type 1, E2 - OSPF external type 2
       i - IS-IS, su - IS-IS summary, L1 - IS-IS level-1, L2 - IS-IS level-2
       ia - IS-IS inter area, * - candidate default, U - per-user static route
       o - ODR, P - periodic downloaded static route

Gateway of last resort is not set

C    192.168.5.0/24 is directly connected, Serial1/1
C    192.168.2.0/24 is directly connected, FastEthernet0/0
```

第三步：查看 Router-B 的路由表。

```
Router-B#sh ip route
Codes: C - connected, S - static, R - RIP, M - mobile, B - BGP
       D - EIGRP, EX - EIGRP external, O - OSPF, IA - OSPF inter area
       N1 - OSPF NSSA external type 1, N2 - OSPF NSSA external type 2
       E1 - OSPF external type 1, E2 - OSPF external type 2
       i - IS-IS, su - IS-IS summary, L1 - IS-IS level-1, L2 - IS-IS level-2
       ia - IS-IS inter area, * - candidate default, U - per-user static route
       o - ODR, P - periodic downloaded static route

Gateway of last resort is not set

C    192.168.5.0/24 is directly connected, Serial1/1
C    192.168.2.0/24 is directly connected, FastEthernet0/0
```

第四步：查看 Router-C 的路由表。

```
Router-C#sh ip route
Codes: C - connected, S - static, R - RIP, M - mobile, B - BGP
       D - EIGRP, EX - EIGRP external, O - OSPF, IA - OSPF inter area
       N1 - OSPF NSSA external type 1, N2 - OSPF NSSA external type 2
       E1 - OSPF external type 1, E2 - OSPF external type 2
       i - IS-IS, su - IS-IS summary, L1 - IS-IS level-1, L2 - IS-IS level-2
       ia - IS-IS inter area, * - candidate default, U - per-user static route
       o - ODR, P - periodic downloaded static route

Gateway of last resort is not set

C    192.168.5.0/24          is directly connected, Serial1/0
C    192.168.3.0/24          is directly connected, Ethernet0/0
```

第五步：在 Router-A 上 ping 路由器 C。

```
Router-A#ping 192.168.2.2
Type escape sequence to abort.
Sending 5, 100-byte ICMP Echos to 192.168.2.2, timeout is 2 seconds:
.....
Success rate is 0 percent (0/5)
```

第六步：在 Router-A 上配置静态路由。

```
Router-A#config
Router-A_config#ip route 192.168.2.0 255.255.255.0 192.168.5.2
Router-A_config#ip route 192.168.3.0 255.255.255.0 192.168.5.2
```

第七步：查看路由表。

```
Router-A#sh ip route
Codes: C - connected, S - static, R - RIP, M - mobile, B - BGP
       D - EIGRP, EX - EIGRP external, O - OSPF, IA - OSPF inter area
```

```
          N1 - OSPF NSSA external type 1, N2 - OSPF NSSA external type 2
          E1 - OSPF external type 1, E2 - OSPF external type 2
          i - IS-IS, su - IS-IS summary, L1 - IS-IS level-1, L2 - IS-IS level-2
          ia - IS-IS inter area, * - candidate default, U - per-user static route
          o - ODR, P - periodic downloaded static route

Gateway of last resort is not set

C       192.168.5.0/24 is directly connected, Serial1/1
C       192.168.0.0/24 is directly connected, FastEthernet0/0
S       192.168.2.0/24 [1/0] via 192.168.5.2
S       192.168.3.0/24 [1/0] via 192.168.5.2
```

第八步：配置 Router-B 的静态路由并查看路由表。

```
Router-B#config terminal
Router-B(config)#ip route 192.168.0.0 255.255.255.0 192.168.5.1
Router-B(config)#ip route 192.168.3.0 255.255.255.0 192.168.2.2
Router-B(config)#^Z
Router-B#show ip route
Codes: C - connected, S - static, R - RIP, M - mobile, B - BGP
          D - EIGRP, EX - EIGRP external, O - OSPF, IA - OSPF inter area
          N1 - OSPF NSSA external type 1, N2 - OSPF NSSA external type 2
          E1 - OSPF external type 1, E2 - OSPF external type 2
          i - IS-IS, su - IS-IS summary, L1 - IS-IS level-1, L2 - IS-IS level-2
          ia - IS-IS inter area, * - candidate default, U - per-user static route
          o - ODR, P - periodic downloaded static route

Gateway of last resort is not set

S       192.168.0.0/24          [1,0] via 192.168.5.1
C       192.168.5.0/24          is directly connected, Serial1/0
C       192.168.2.0/24          is directly connected, FastEthernet0/0
S       192.168.3.0/24          [1,0] via 192.168.2.2
```

第九步：配置 Router-C 的静态路由并查看路由表。

```
Router-C#config terminal
Router-C(config)#ip route 192.168.0.0 255.255.0.0 192.168.2.1
Router-C(config)#^Z
Router-C#show ip route
Codes: C - connected, S - static, R - RIP, M - mobile, B - BGP
          D - EIGRP, EX - EIGRP external, O - OSPF, IA - OSPF inter area
          N1 - OSPF NSSA external type 1, N2 - OSPF NSSA external type 2
          E1 - OSPF external type 1, E2 - OSPF external type 2
          i - IS-IS, su - IS-IS summary, L1 - IS-IS level-1, L2 - IS-IS level-2
          ia - IS-IS inter area, * - candidate default, U - per-user static route
          o - ODR, P - periodic downloaded static route

Gateway of last resort is not set
```

S	192.168.0.0/16	[1,0] via 192.168.2.1	
C	192.168.2.0/24	is directly connected,	FastEthernet0/0
C	192.168.3.0/24	is directly connected,	Ethernet1/0

第十步：测试。

```
Router-C#ping 192.168.0.1
Type escape sequence to abort.
Sending 5, 100-byte ICMP Echos to 192.168.0.1, timeout is 2 seconds:
!!!!!
Success rate is 100 percent (5/5), round-trip min/avg/max = 40/52/72 ms
```

七、注意事项和排错

1. 非直连的网段都要配置路由。
2. 以太网接口只有连接到主机或交换机才能变为 up 状态。
3. 串口注意 DCE 和 DTE 的问题。

八、配置序列

Router-B 的序列：

```
Building configuration...
Current configuration : 941 bytes
!
version 12.4
service timestamps debug datetime msec
service timestamps log datetime msec
no service password-encryption
!
hostname Router-B
!
boot-start-marker
boot-end-marker
!
!no aaa new-model
memory-size iomem 5
!
ip cef
no ip domain lookup
!
interface FastEthernet0/0
 ip address 192.168.2.1 255.255.255.0
 duplex auto
 speed auto
```

```
!
interface Serial1/0
  no ip address
  serial restart-delay 0
!
interface Serial1/1
  ip address 192.168.5.2 255.255.255.0
  serial restart-delay 0
!
no ip http server
no ip http secure-server
ip route 192.168.0.0 255.255.255.0 192.168.5.1
ip route 192.168.3.0 255.255.255.0 192.168.2.2
```

九、思考题

1. 什么情况下可以采用 Router-C 的超网配置方法？
2. 为什么只有当所有路由器都配置了路由以后才能实现通信？
3. 静态路由有什么优势？在什么情况下使用？

实训项目八　多路由器间静态路由配置

一、实训目的

  1. 学习如何在包含多个路由器的网络环境中配置静态路由，以实现不同网络段之间的有效通信。

  2. 掌握在多路由器网络中进行网络规划的技能，包括地址分配、路由策略选择等。

  3. 学习如何配置路由器之间的互联接口，包括 IP 地址分配和接口类型设置。

二、实训设备

(1) 二层交换机(S-A 和 S-B) 2 台。
(2) 路由器(R-C 和 R-D) 2 台。
(3) PC 4 台。
(4) 网线 7 根。

三、实训拓扑

  该实验拓扑结构如图 2-14 所示。

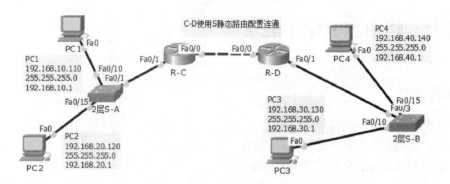

图 2-14 实验拓扑图

四、实训要求

1. 按照图 2-14 所示进行设备连线，并对各设备进行以下配置。
(1) 交换机 A 划分 vl 10(9~12)、vl 20(13~16)，Trunk (1)。
(2) 交换机 B 划分 vl 30(17~20)、vl 40(21~24)，Trunk (3)。
(3) 设置路由器 C：

E 0/1.10	192.168.10.1/24	封装 vl 10;
E 0/1.20	192.168.20.1/24	封装 vl 20;
F 0/0	192.168.100.1/30	

(4) 设置路由器 D：

F 0/1.30	192.168.30.1/24	封装 vl 30;
F 0/1.40	192.168.40.1/24	封装 vl 40;
F 0/0	192.168.100.2/30	

(5) 配置 RC 的静态路由：

RC_config#ip route 192.168.30.0 255.255.255.0 192.168.100.2
RC_config#ip route 192.168.40.0 255.255.255.0 192.168.100.2

(6) 配置 RD 的静态路由：

RD_config#ip route 192.168.10.0 255.255.255.0 192.168.100.1
RD_config#ip route 192.168.20.0 255.255.255.0 192.168.100.1

(7) 根据连线位置正确配置 PC1、PC2 的地址(注意配上网关地址以及与 VLAN 的对应关系)。

2. 实验结果：PC1---ping---PC2、3、4 通。
3. 查看设备状态：show vlan   show run   show ip route。

实验七 路由器 RIP-1 配置

一、实验目的

1. 掌握 RIP 的基本原理和运作方式。
2. 学习如何在路由器上配置 RIP-1 协议。
3. 诊断和排除 RIP 配置过程中的故障问题。

二、相关知识

RIP 是一种距离矢量路由协议，它使用跳数(hop count)作为路由度量来选择最佳路径。RIP-1 是 RIP 第一个版本，它不支持 VLSM(变长子网掩码)和 CIDR(无类别域间路由)。RIP-1 使用广播方式发送路由更新信息，并且 RIP 路由器每隔 30 秒发送一次路由更新。

路由器 RIP-1 配置主要适用于以下场景。

(1) 中小型网络。RIP-1 适用于小型到中型网络，在这些网络中，跳数作为度量标准通常能够满足需求。

(2) 不支持 VLSM 的网络。在不支持 VLSM 和 CIDR 的网络环境中，RIP-1 可能是唯一可用的路由协议。

三、实验设备

1. DCR-1751 3 台。
2. CR-V35FC 1 条。
3. CR-V35MT 1 条。

四、实验拓扑

该实验拓扑结构如图 2-15 所示。

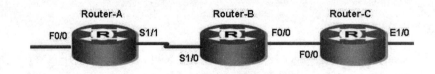

图 2-15　实验拓扑图

五、实验要求

实验中各设备接口的配置如表 2-6 所示。

表2-6 设备接口配置

Router-A		Router-B		Router-C	
S1/1(DCE)	192.168.5.1	S1/0(DTE)	192.168.5.2	F0/0	192.168.2.2
F0/0	192.168.0.1	F0/0	192.168.2.1	E1/0	192.168.3.1

六、实验步骤

第一步：按照表 2-6 配置所有接口的 IP 地址，确保所有接口均处于 up 状态，并测试网络连通性。

第二步：查看 Router-A 的路由表。

```
Router-A#sh ip route
Codes: C - connected, S - static, R - RIP, M - mobile, B - BGP
       D - EIGRP, EX - EIGRP external, O - OSPF, IA - OSPF inter area
       N1 - OSPF NSSA external type 1, N2 - OSPF NSSA external type 2
       E1 - OSPF external type 1, E2 - OSPF external type 2
       i - IS-IS, su - IS-IS summary, L1 - IS-IS level-1, L2 - IS-IS level-2
       ia - IS-IS inter area, * - candidate default, U - per-user static route
       o - ODR, P - periodic downloaded static route

Gateway of last resort is not set

C    192.168.5.0/24 is directly connected, Serial1/1
C    192.168.0.0/24 is directly connected, FastEthernet0/0
```

第三步：查看 Router-B 的路由表。

```
Router-B#sh ip route
Codes: C - connected, S - static, R - RIP, M - mobile, B - BGP
       D - EIGRP, EX - EIGRP external, O - OSPF, IA - OSPF inter area
       N1 - OSPF NSSA external type 1, N2 - OSPF NSSA external type 2
       E1 - OSPF external type 1, E2 - OSPF external type 2
       i - IS-IS, su - IS-IS summary, L1 - IS-IS level-1, L2 - IS-IS level-2
       ia - IS-IS inter area, * - candidate default, U - per-user static route
       o - ODR, P - periodic downloaded static route

Gateway of last resort is not set

C    192.168.5.0/24 is directly connected, Serial1/1
C    192.168.2.0/24 is directly connected, FastEthernet0/0
```

第四步：查看 Router-C 的路由表。

```
Router-C#sh ip route
Codes: C - connected, S - static, R - RIP, M - mobile, B - BGP
       D - EIGRP, EX - EIGRP external, O - OSPF, IA - OSPF inter area
```

　　　　　　N1 - OSPF NSSA external type 1, N2 - OSPF NSSA external type 2
　　　　　　E1 - OSPF external type 1, E2 - OSPF external type 2
　　　　　　i - IS-IS, su - IS-IS summary, L1 - IS-IS level-1, L2 - IS-IS level-2
　　　　　　ia - IS-IS inter area, * - candidate default, U - per-user static route
　　　　　　o - ODR, P - periodic downloaded static route

Gateway of last resort is not set
C　　192.168.2.0/24 is directly connected, FastEthernet0/0
C　　192.168.3.0/24 is directly connected, FastEthernet1/0

第五步：在 Router-A 上 ping 路由器 C。

Router-A#ping 192.168.2.2
Type escape sequence to abort.
Sending 5, 100-byte ICMP Echos to 192.168.2.2, timeout is 2 seconds:
.....
Success rate is 0 percent (0/5)　　　　　　　　　　　！不通

第六步：在 Router-A 上配置 RIP 协议并查看路由表。

Router-A(config)#router rip　　　　　　　　　　　！启动 RIP 协议
Router-A(config-router)#network 192.168.0.0　　　！宣告网段
Router-A(config-router)#network 192.168.5.0
Router-A(config-router)#^Z
Router-A#sh ip route
Codes: C - connected, S - static, R - RIP, M - mobile, B - BGP
　　　　D - EIGRP, EX - EIGRP external, O - OSPF, IA - OSPF inter area
　　　　N1 - OSPF NSSA external type 1, N2 - OSPF NSSA external type 2
　　　　E1 - OSPF external type 1, E2 - OSPF external type 2
　　　　i - IS-IS, su - IS-IS summary, L1 - IS-IS level-1, L2 - IS-IS level-2
　　　　ia - IS-IS inter area, * - candidate default, U - per-user static route
　　　　o - ODR, P - periodic downloaded static route

Gateway of last resort is not set
C　　192.168.5.0/24 is directly connected, Serial1/1
C　　192.168.0.0/24 is directly connected, FastEthernet0/0

注意到 RIP 学习到的路由未出现。

第七步：在 Router-B 上配置 RIP 协议并查看路由表。

Router-B(config)#router rip
Router-B(config-router)#network 192.168.5.0
Router-B(config-router)#network 192.168.2.0
Router-B(config-router)#^Z

```
Router-B#sh ip route
Codes: C - connected, S - static, R - RIP, M - mobile, B - BGP
       D - EIGRP, EX - EIGRP external, O - OSPF, IA - OSPF inter area
       N1 - OSPF NSSA external type 1, N2 - OSPF NSSA external type 2
       E1 - OSPF external type 1, E2 - OSPF external type 2
       i - IS-IS, su - IS-IS summary, L1 - IS-IS level-1, L2 - IS-IS level-2
       ia - IS-IS inter area, * - candidate default, U - per-user static route
       o - ODR, P - periodic downloaded static route

Gateway of last resort is not set

C    192.168.5.0/24 is directly connected, Serial1/1
R    192.168.0.0/24 [120/1] via 192.168.5.1, 00:00:27, Serial1/1     ！从 A 学习到的路由
C    192.168.2.0/24 is directly connected, FastEthernet0/0
```

第八步：在 Router-C 上配置 RIP 协议并查看路由表。

```
Router-C(config)#router rip
Router-C(config-router)#network 192.168.2.0
Router-C(config-router)#network 192.168.3.0
Router-C(config-router)#^Z
Router-C#sh ip route
Codes: C - connected, S - static, R - RIP, M - mobile, B - BGP
       D - EIGRP, EX - EIGRP external, O - OSPF, IA - OSPF inter area
       N1 - OSPF NSSA external type 1, N2 - OSPF NSSA external type 2
       E1 - OSPF external type 1, E2 - OSPF external type 2
       i - IS-IS, su - IS-IS summary, L1 - IS-IS level-1, L2 - IS-IS level-2
       ia - IS-IS inter area, * - candidate default, U - per-user static route
       o - ODR, P - periodic downloaded static route

Gateway of last resort is not set

R    192.168.5.0/24 [120/1] via 192.168.2.1, 00:00:08, FastEthernet0/0
R    192.168.0.0/24 [120/1] via 192.168.2.1, 00:00:08, FastEthernet0/0
C    192.168.2.0/24 is directly connected, FastEthernet0/0
C    192.168.3.0/24 is directly connected, FastEthernet1/0
```

第九步：再次查看 Router-A 和 Router-B 的路由表。

```
Router-B#sh ip route
Codes: C - connected, S - static, R - RIP, M - mobile, B - BGP
       D - EIGRP, EX - EIGRP external, O - OSPF, IA - OSPF inter area
       N1 - OSPF NSSA external type 1, N2 - OSPF NSSA external type 2
       E1 - OSPF external type 1, E2 - OSPF external type 2
       i - IS-IS, su - IS-IS summary, L1 - IS-IS level-1, L2 - IS-IS level-2
```

ia - IS-IS inter area, * - candidate default, U - per-user static route
o - ODR, P - periodic downloaded static route

Gateway of last resort is not set

C 192.168.5.0/24 is directly connected, Serial1/1
R 192.168.0.0/24 [120/1] via 192.168.5.1, 00:00:21, Serial1/1
C 192.168.2.0/24 is directly connected, FastEthernet0/0
R 192.168.3.0/24 [120/1] via 192.168.2.2, 00:00:19, FastEthernet0/0

Router-A#sh ip route
Codes: C - connected, S - static, R - RIP, M - mobile, B - BGP
   D - EIGRP, EX - EIGRP external, O - OSPF, IA - OSPF inter area
   N1 - OSPF NSSA external type 1, N2 - OSPF NSSA external type 2
   E1 - OSPF external type 1, E2 - OSPF external type 2
   i - IS-IS, su - IS-IS summary, L1 - IS-IS level-1, L2 - IS-IS level-2
   ia - IS-IS inter area, * - candidate default, U - per-user static route
   o - ODR, P - periodic downloaded static route

Gateway of last resort is not set

C 192.168.5.0/24 is directly connected, Serial1/1
C 192.168.0.0/24 is directly connected, FastEthernet0/0
R 192.168.2.0/24 [120/1] via 192.168.5.2, 00:00:17, Serial1/1
R 192.168.3.0/24 [120/1] via 192.168.5.2, 00:00:17, Serial1/1
！注意到所有网段都学习到了路由

第十步：相关的查看命令。

Router-A#sh ip protocols   ！显示协议细节
Routing Protocol is "rip"
 Outgoing update filter list for all interfaces is not set
 Incoming update filter list for all interfaces is not set
 Sending updates every 30 seconds, next due in 17 seconds ！注意定时器的值
 Invalid after 180 seconds, hold down 180, flushed after 240
 Redistributing: rip
 Default version control: send version 1, receive any version
  Interface   Send Recv Triggered RIP Key-chain
  FastEthernet0/0   1 1 2
  Serial1/1   1 1 2
 Automatic network summarization is in effect
 Maximum path: 4
 Routing for Networks:
  192.168.0.0
  192.168.5.0
 Routing Information Sources:

Gateway	Distance	Last Update
192.168.5.2	120	00:00:14

Distance: (default is 120)　　　　　　　　　　　　！注意默认的管理距离

Router-A#show ip rip database　　　　　　　　　！显示 RIP 数据库
192.168.0.0/24　　　auto-summary
192.168.0.0/24　　　directly connected, FastEthernet0/0
192.168.2.0/24　　　auto-summary
192.168.2.0/24
　　　[1] via 192.168.5.2, 00:00:11, Serial1/1　　！收到 RIP 广播的时间
192.168.3.0/24　　　auto-summary
192.168.3.0/24
　　　[1] via 192.168.5.2, 00:00:11, Serial1/1
192.168.5.0/24　　　auto-summary
192.168.5.0/24　　　directly connected, Serial1/1

Router-A#sh ip route rip　　　　　　　　　　　　！仅显示 RIP 学习到的路由
R　　192.168.2.0/24 [120/1] via 192.168.5.2, 00:00:01, Serial1/1
R　　192.168.3.0/24 [120/1] via 192.168.5.2, 00:00:01, Serial1/1

七、注意事项和排错

1. 只能宣告直接连接的网段。
2. 宣告时不附加子网掩码。
3. 分配地址时最好使用连续的子网，以避免 RIP 汇聚时出现错误。

八、配置序列

Router-B 的序列：

Building configuration...

Current configuration : 902 bytes
!
version 12.4
service timestamps debug datetime msec
service timestamps log datetime msec
no service password-encryption
!
hostname Router-B
!
boot-start-marker
boot-end-marker

```
!
no aaa new-model
memory-size iomem 5
!
ip cef
no ip domain lookup
!
interface FastEthernet0/0
  ip address 192.168.2.1 255.255.255.0
  duplex auto
  speed auto
!
interface Serial1/0
  no ip address
  serial restart-delay 0
!
interface Serial1/1
  ip address 192.168.5.2 255.255.255.0
  serial restart-delay 0
!
interface Serial1/2
  no ip address
  shutdown
  serial restart-delay 0
!
interface Serial1/3
  no ip address
  shutdown
  serial restart-delay 0
!
router rip
  network 192.168.2.0
  network 192.168.5.0
```

九、思考题

1. 为什么当 Router-B 没有配置 RIP 协议时，Router-A 的路由表中没有出现 RIP 路由？
2. 如果使用非连续的子网，会导致什么结果？
3. RIP 的广播周期是多少？

实训项目九　多路由器间的 RIP-1 路由配置

一、实训目的

1. 深入理解 RIP 的工作原理，包括路由更新、路由选择和路由收敛等过程。
2. 学会在多个路由器上配置 RIP-1，包括网络宣告和版本选择等配置。
3. 能够利用 RIP-1 协议实现不同网络之间的互联，确保网络层的连通性。

二、实训设备

(1) 二层交换机(S-A 和 S-B) 2 台。
(2) 路由器(R-C 和 R-D) 2 台。
(3) PC 机 4 台。
(4) 网线 7 根。

三、实训拓扑

该实验拓扑结构如图 2-16 所示。

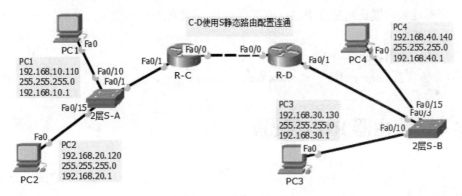

图 2-16　实验拓扑图

四、实训要求

参照图 2-16 所示进行连线，各个设备做如下配置。

(1) 交换机 A 划分 vl 10(9~12)、vl 20(13~16)，Trunk (1)。
(2) 交换机 B 划分 vl 30(17~20)、vl 40(21~24)，Trunk (3)。
(3) 设置路由器 C：

E 0/1.10	192.168.10.1/24	封装 vl 10;
E 0/1.20	192.168.20.1/24	封装 vl 20;
F 0/0	192.168.100.1/30	

(4) 设置路由器 D:

F 0/1.30	192.168.30.1/24	封装 vl 30;
F 0/1.40	192.168.40.1/24	封装 vl 40;
F 0/0	192.168.100.2/30	

(5) 启用路由器的动态 RIP 协议，配置路由器 C 的动态路由 RIP：

RC(config)#router rip
RC(config-router)#network 192.168.10.0
RC(config-router)#network 192.168.20.0
RC(config-router)#network 192.168.100.0

(6) 配置路由器 D 的动态路由 RIP：

RD(config)#router rip
RD(config-router)#network 192.168.30.0
RD(config-router)#network 192.168.40.0
RD(config-router)#network 192.168.100.0

验证：sh ip rip   sh ip route。

(7) 根据连线位置正确配置 PC1、PC2 的地址(注意配置网关地址，并确保与 VLAN 的对应关系)。

实验结果：PC1---ping---PC2、3、4   通。

查看设备状态：show vlan   show run   show ip route。

实验八　路由器 RIP-2 配置

一、实验目的

1. 掌握 RIP-2 (RIP 版本 2)的基本原理和运作方式，以及与 RIP 版本 1(RIP-1)的区别。
2. 理解 RIP-2 如何支持 VLSM(变长子网掩码)和 CIDR(无类别域间路由)。
3. 学会在路由器上配置 RIP-2，包括网络宣告、版本选择、认证和路由汇总。

二、相关知识

RIP-2 支持 VLSM 和 CIDR，这使得它在更复杂的网络环境中更具优势。与 RIP-1 不同，RIP-2 使用点播而不是广播来发送路由更新，从而减少了网络上的不必要流量。与 RIP-1 相同，RIP-2 的路由更新计时器默认每 30 秒发送一次更新。

路由器 RIP-2 配置主要应用于以下场景。

(1) 中小型企业网络。RIP-2 适用中小型企业网络，特别是那些需要支持 VLSM 和 CIDR

的网络。

(2) 简单园区网络。对于简单的园区网络，RIP-2 可以提供稳定且易于管理的路由解决方案。

(3) 多 VLAN 环境。在需要跨越多个 VLAN 进行路由的网络环境中，RIP-2 能够有效地工作。

三、实验设备

1. DCR-1751 3 台。
2. CR-V35FC 1 条。
3. CR-V35MT 1 条。

四、实验拓扑

该实验拓扑结构如图 2-17 所示。

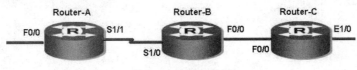

图 2-17　实验拓扑图

五、实验要求

实验中各设备的接口配置如表 2-7 所示。

表 2-7　设备接口配置

Router-A		Router-B		Router-C	
S1/1(DCE)	192.168.5.1/24	S1/0(DTE)	192.168.5.2/24	F0/0	192.168.2.2/24
F0/0	172.16.3.1/24	F0/0	192.168.2.1/24	E1/0	172.16.4.1/24

六、实验步骤

第一步：参照表 2-7 配置各设备所有接口的 IP 地址，确保所有接口均处于 up 状态，并测试网络连通性。

第二步：配置 Router-A 的 RIP-1 协议。

```
Router-A(config)#router rip
Router-A(config-router)# network 192.168.5.0
Router-A(config-router)#network   172.16.3.0 255.255.255.0   !在 RIP-1 里掩码是没有意义的
Router-A(config-router)
#^Z
```

第三步：查看 Router-A 的路由表。

```
Router-A#sh ip route
Codes: C - connected, S - static, R - RIP, M - mobile, B - BGP
       D - EIGRP, EX - EIGRP external, O - OSPF, IA - OSPF inter area
       N1 - OSPF NSSA external type 1, N2 - OSPF NSSA external type 2
       E1 - OSPF external type 1, E2 - OSPF external type 2
       i - IS-IS, su - IS-IS summary, L1 - IS-IS level-1, L2 - IS-IS level-2
       ia - IS-IS inter area, * - candidate default, U - per-user static route
       o - ODR, P - periodic downloaded static route

Gateway of last resort is not set

     172.16.0.0/24   is subnetted, 1 subnets
C    172.16.3.0    is directly connected, FastEthernet0/0      ！直连的路由
C    192.168.5.0/24 is directly connected, Serial1/1           ！直连的路由
```

第四步：配置 Router-B 的 RIP-1 协议。

```
Router-B(config)#router rip
Router-B(config-router)#network 192.168.5.0
Router-B(config-router)#network 192.168.2.0
Router-B(config-router)#^Z
```

第五步：查看 Router-B 的路由表。

```
Router-B#show ip route
Codes: C - connected, S - static, R - RIP, M - mobile, B - BGP
       D - EIGRP, EX - EIGRP external, O - OSPF, IA - OSPF inter area
       N1 - OSPF NSSA external type 1, N2 - OSPF NSSA external type 2
       E1 - OSPF external type 1, E2 - OSPF external type 2
       i - IS-IS, su - IS-IS summary, L1 - IS-IS level-1, L2 - IS-IS level-2
       ia - IS-IS inter area, * - candidate default, U - per-user static route
       o - ODR, P - periodic downloaded static route

Gateway of last resort is not set

R    172.16.0.0/16   [120/1]  via 192.168.5.1, 00:00:06, Serial1/1   ！注意是有类的地址
C    192.168.5.0/24  is directly connected, Serial1/1
C    192.168.2.0/24  is directly connected, FastEthernet0/0
```

第六步：配置 Router-C 的 RIP-1 协议。

```
Router-C(config)#router rip
Router-C(config-router)#network 192.168.2.0
Router-C(config-router)#network 172.16.4.0 255.255.255.0     ！RIP-1 是有类的路由协议
Router-C_config_rip#^Z
```

第七步：再次查看 Router-B 的路由表。

```
Router-B#sh ip route
```

Codes: C - connected, S - static, R - RIP, M - mobile, B - BGP
       D - EIGRP, EX - EIGRP external, O - OSPF, IA - OSPF inter area
       N1 - OSPF NSSA external type 1, N2 - OSPF NSSA external type 2
       E1 - OSPF external type 1, E2 - OSPF external type 2
       i - IS-IS, su - IS-IS summary, L1 - IS-IS level-1, L2 - IS-IS level-2
       ia - IS-IS inter area, * - candidate default, U - per-user static route
       o - ODR, P - periodic downloaded static route

Gateway of last resort is not set

R   172.16.0.0/16 [120/1] via 192.168.5.1, 00:00:01, Serial1/1
                    [120/1] via 192.168.2.2, 00:00:02, FastEthernet0/0
                    ！由于有类路由的自动汇总，出现了错误的路由

C   192.168.5.0/24 is directly connected, Serial1/1
C   192.168.2.0/24 is directly connected, FastEthernet0/0

第八步：在所有路由器上配置 RIP-2 协议并关闭自动汇总。

Router-C(config)#router rip
Router-C(config-router)#version 2   ！指明为版本 2
Router-C(config-router)#no auto-summary   ！关闭自动汇总

Router-B(config)#router rip
Router-B(config-router)#version 2
Router-B(config-router)#no auto-summary

Router-A(config)#router rip
Router-A(config-router)#version 2
Router-A(config-router)#no auto-summary

第九步：再次查看所有的路由表。
Router-B 的路由表：

Router-B#sh ip route
Codes: C - connected, S - static, R - RIP, M - mobile, B - BGP
       D - EIGRP, EX - EIGRP external, O - OSPF, IA - OSPF inter area
       N1 - OSPF NSSA external type 1, N2 - OSPF NSSA external type 2
       E1 - OSPF external type 1, E2 - OSPF external type 2
       i - IS-IS, su - IS-IS summary, L1 - IS-IS level-1, L2 - IS-IS level-2
       ia - IS-IS inter area, * - candidate default, U - per-user static route
       o - ODR, P - periodic downloaded static route

Gateway of last resort is not set

      172.16.0.0/16 is variably subnetted, 3 subnets, 2 masks
R   172.16.4.0/24 [120/1] via 192.168.2.2, 00:00:12, FastEthernet0/0   ！正确的路由
R   172.16.0.0/16 [120/1] via 192.168.5.1, 00:01:36, Serial1/1
R   172.16.3.0/24 [120/1] via 192.168.5.1, 00:00:11, Serial1/1

```
C    192.168.5.0/24 is directly connected, Serial1/1
C    192.168.2.0/24 is directly connected, FastEthernet0/0
```

Router-A 的路由表：

```
Router-A#sh ip route
Codes: C - connected, S - static, R - RIP, M - mobile, B - BGP
       D - EIGRP, EX - EIGRP external, O - OSPF, IA - OSPF inter area
       N1 - OSPF NSSA external type 1, N2 - OSPF NSSA external type 2
       E1 - OSPF external type 1, E2 - OSPF external type 2
       i - IS-IS, su - IS-IS summary, L1 - IS-IS level-1, L2 - IS-IS level-2
       ia - IS-IS inter area, * - candidate default, U - per-user static route
       o - ODR, P - periodic downloaded static route

Gateway of last resort is not set

     172.16.0.0/24 is subnetted, 2 subnets
R       172.16.4.0 [120/2] via 192.168.5.2, 00:00:01, Serial1/1        ！正确的路由
C       172.16.3.0 is directly connected, FastEthernet0/0
C    192.168.5.0/24 is directly connected, Serial1/1
R    192.168.2.0/24 [120/1] via 192.168.5.2, 00:00:01, Serial1/1

Router-C#sh ip route
Codes: C - connected, S - static, R - RIP, M - mobile, B - BGP
       D - EIGRP, EX - EIGRP external, O - OSPF, IA - OSPF inter area
       N1 - OSPF NSSA external type 1, N2 - OSPF NSSA external type 2
       E1 - OSPF external type 1, E2 - OSPF external type 2
       i - IS-IS, su - IS-IS summary, L1 - IS-IS level-1, L2 - IS-IS level-2
       ia - IS-IS inter area, * - candidate default, U - per-user static route
       o - ODR, P - periodic downloaded static route

Gateway of last resort is not set

     172.16.0.0/24 is subnetted, 2 subnets
C       172.16.4.0 is directly connected, FastEthernet1/0
R       172.16.3.0 [120/2] via 192.168.2.1, 00:00:19, FastEthernet0/0
R    192.168.5.0/24 [120/1] via 192.168.2.1, 00:00:19, FastEthernet0/0
C    192.168.2.0/24 is directly connected, FastEthernet0/0
```
！注意到所有网段都学习到了正确掩码的路由

第十步：相关的查看命令。

```
Router-A#show ip rip protocol          ！显示协议细节
….
Router-A#show ip rip database          ！显示 RIP 数据库
….
Router-A#show ip route rip             ！仅显示 RIP 学习到的路由
….
```

七、注意事项和排错

1. 只能宣告直接连接的网段。
2. 宣告时不附加子网掩码。
3. 分配地址时最好使用连续的子网，以避免 RIP 汇聚时出现错误。

八、配置序列

Router-B 的序列：

```
Building configuration...
Current configuration : 930 bytes
!
version 12.4
service timestamps debug datetime msec
service timestamps log datetime msec
no service password-encryption
!
hostname Router-B
!
boot-start-marker
boot-end-marker
!
no aaa new-model
memory-size iomem 5
!
ip cef
no ip domain lookup
!
interface FastEthernet0/0
 ip address 192.168.2.1 255.255.255.0
 duplex auto
 speed auto
!
interface Serial1/0
 no ip address
 serial restart-delay 0
!
interface Serial1/1
 ip address 192.168.5.2 255.255.255.0
 serial restart-delay 0
!
interface Serial1/2
 no ip address
```

```
    shutdown
    serial restart-delay 0
!
interface Serial1/3
    no ip address
    shutdown
    serial restart-delay 0
!
router rip
    version 2
    network 192.168.2.0
    network 192.168.5.0
    no auto-summary
```

九、思考题

1. RIP-1 与 RIP-2 的区别是什么？
2. 如果子网不是连续的，会出现什么问题？
3. RIP-2 的组播地址是什么(可以使用 debug ip rip protocol 命令查看，但要注意使用 no debug all 命令关闭调试)？

实训项目十　多路由器间的 RIP-2 路由配置

一、实训目的

1. 理解 RIP-2 协议的工作原理。
2. 学会在多路由器环境中配置和优化 RIP-2 协议。
3. 通过 RIP-2 协议实现复杂网络中不同子网和 VLAN 之间的互联。

二、实训设备

(1) 二层交换机(S-A 和 S-B) 2 台。
(2) 路由器(R-C 和 R-D) 2 台。
(3) PC 4 台。
(4) 网线 7 根。

三、实训拓扑

该实验拓扑结构如图 2-18 所示。

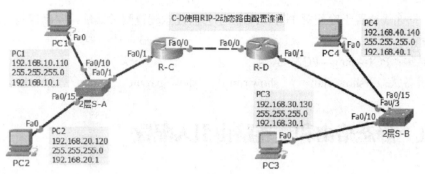

图 2-18　实验拓扑图

四、实训要求

按照图 2-18 所示进行设备连线，并对各设备进行以下配置。

(1) 交换机 A 划分 vl 10(9~12)、vl 20(13~16)，Trunk (1)。
(2) 交换机 B 划分 vl 30(17~20)、vl 40(21~24)，Trunk (3)。
(3) 设置路由器 C：

E 0/1.10	192.168.10.1/24	封装 vl 10；
E 0/1.20	192.168.20.1/24	封装 vl 20；
F 0/0	192.168.100.1/30	

(4) 设置路由器 D：

F 0/1.30	192.168.30.1/24	封装 vl 30；
F 0/1.40	192.168.40.1/24	封装 vl 40；
F 0/0	192.168.100.2/30	

(5) 启用路由器的动态 RIP 协议，配置路由器 C 的动态路由 RIP：

RC(config)#router rip
RC(config)#version 2
RC(config-router)#network 192.168.10.0
RC(config-router)#network 192.168.20.0
RC(config-router)#network 192.168.100.0

(6) 配置路由器 D 的动态路由 RIP：

RD(config)#router rip
RD(config-router)#version 2
RD(config-router)#network 192.168.30.0
RD(config-router)p#network 192.168.40.0
RD(config-router)p#network 192.168.100.0

验证：sh ip rip　　　　sh ip route。

(7) 根据连线位置正确配置 PC1、PC2 的地址(注意配置网关地址，并确保与 VLAN 的对应关系)。

实验结果：PC1---ping---PC2、3、4　　　通。

查看设备状态：show vlan　　　show run　　　show ip route。

实验九　静态路由和直连路由引入配置

一、实验目的

1. 掌握静态路由、直连路由(Direct Route)和引入路由(Imported Route)的概念和区别。
2. 学习配置静态路由和直连路由，并了解如何引入外部路由信息。

二、相关知识

静态路由是由网络管理员手动配置的路由信息，除非管理员修改配置，否则它不会自动发生变化。直连路由是指路由器接口直接连接的网络路径，路由器自动将直连网络添加到路由表中，无需手动配置。引入路由则是指将外部路由信息(例如从其他路由协议或自治系统)引入到当前路由器的路由表中。

本实验主要应用于以下场景。

(1) 静态路由主要应用于小型网络中，由于网络结构简单，使用静态路由可以轻松实现网络互连。当需要精确控制数据包的传输路径时，静态路由非常有用。

(2) 直连路由是所有网络的基本组成部分，因为它直接关联到路由器的物理接口。

(3) 在运行多个路由协议的网络中，可能需要将一个协议的路由信息引入到另一个协议中。

三、实验设备

DCR-1751 2 台。

四、实验拓扑

该实验拓扑结构如图 2-19 所示。

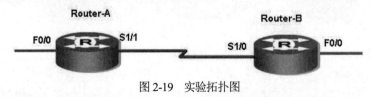

图 2-19　实验拓扑图

五、实验要求

实验中各设备的接口配置如表 2-8 所示。

表 2-8 设备接口配置

Router-A		Router-B	
S1/1(DCE)	192.168.5.1/24	S1/0	192.168.5.2/24
F0/0	192.168.0.1/24	F0/0	192.168.2.1/24

六、实验步骤

1. 引入到 RIP 协议中

第一步：参照表 2-8 配置路由器的所有接口地址并测试网络连通性。

第二步：配置 Router-A 的静态路由，查看直连和静态路由。

```
Router-A#config
Router-A(config)#ip route 191.13.2.0 255.255.255.0 192.168.0.4    ！配置静态路由
Router-A(config)#^Z
Router-A#sh ip route
Codes: C - connected, S - static, R - RIP, M - mobile, B - BGP
       D - EIGRP, EX - EIGRP external, O - OSPF, IA - OSPF inter area
       N1 - OSPF NSSA external type 1, N2 - OSPF NSSA external type 2
       E1 - OSPF external type 1, E2 - OSPF external type 2
       i - IS-IS, su - IS-IS summary, L1 - IS-IS level-1, L2 - IS-IS level-2
       ia - IS-IS inter area, * - candidate default, U - per-user static route
       o - ODR, P - periodic downloaded static route

Gateway of last resort is not set

S       191.13.2.0/24       [1,0] via 192.168.0.4
C       192.168.0.0/24      is directly connected, FastEthernet0/0
C       192.168.5.0/24      is directly connected, Serial1/1
```

第三步：在 Router-A 上配置 RIP 协议，并将直连和静态路由引入。

```
Router-A(config)#router rip
Router-A(config-router)#network 192.168.5.0          ！注意并没有宣告 192.168.0.0
Router-A(config-router)#redistribute connected       ！将直连的路由引入
Router-A(config-router)#redistribute static          ！将静态路由引入
```

第四步：在 Router-B 上配置 RIP 协议，查看从 Router-A 学习到的被引入的路由。

```
Router-B#config   t
Router-B_config#router rip
Router-B_config_rip#network 192.168.5.0
Router-B_config_rip#^Z
Router-B#sh ip route
Codes: C - connected, S - static, R - RIP, M - mobile, B - BGP
       D - EIGRP, EX - EIGRP external, O - OSPF, IA - OSPF inter area
```

```
            N1 - OSPF NSSA external type 1, N2 - OSPF NSSA external type 2
            E1 - OSPF external type 1, E2 - OSPF external type 2
            i - IS-IS, su - IS-IS summary, L1 - IS-IS level-1, L2 - IS-IS level-2
            ia - IS-IS inter area, * - candidate default, U - per-user static route
            o - ODR, P - periodic downloaded static route

Gateway of last resort is not set

R       191.13.0.0/16        [120,1] via 192.168.5.1(on Serial1/0) ！注意是有类的路由
R       192.168.0.0/16       [120,1] via 192.168.5.1(on Serial1/0)
C       192.168.5.0/24       is directly connected, Serial1/0
C       192.168.2.0/24       is directly connected, FastEthernet0/0
```

2. 引入到 OSPF 协议中

第一步和第二步同上。

第三步：在 Router-A 上配置 OSPF 协议，并将直连和静态引入。

```
Router-A#conf  t
Router-A(config)#router ospf 1
Router-A(config-router)#network 192.168.5.0 255.255.255.0 area 0
Router-A(config-router)#redistribute connected subnets
Router-A(config-router)#redistribute static subnets
```

第四步：在 Router-B 上配置 OSPF 协议，并查看从 Router-A 学习到的路由。

```
Router-B#config terminal
Router-B(config)#router ospf 1
Router-B(config-router)#network 192.168.5.0 255.255.255.0 area 0
Router-B(config-router)#exit
Router-B(config)#no router rip
Router-B(config)#^Z
Router-B#sh ip route
Codes: C - connected, S - static, R - RIP, M - mobile, B - BGP
       D - EIGRP, EX - EIGRP external, O - OSPF, IA - OSPF inter area
       N1 - OSPF NSSA external type 1, N2 - OSPF NSSA external type 2
       E1 - OSPF external type 1, E2 - OSPF external type 2
       i - IS-IS, su - IS-IS summary, L1 - IS-IS level-1, L2 - IS-IS level-2
       ia - IS-IS inter area, * - candidate default, U - per-user static route
       o - ODR, P - periodic downloaded static route

Gateway of last resort is not set

       191.13.0.0/24 is subnetted, 1 subnets
O E2    191.13.2.0 [110/20] via 192.168.5.1, 00:00:51, Serial1/1       ！注意管理距离和花费值
C       192.168.5.0/24 is directly connected, Serial1/1
O E2 192.168.0.0/24 [110/20] via 192.168.5.1, 00:04:07, Serial1/1
C       192.168.2.0/24 is directly connected, FastEthernet0/0
```

七、注意事项和排错

1. 注意是将已经存在的路由引入到动态路由协议中,静态路由要先配置。
2. 引入成功后,该动态路由协议将引入的路由发布出去,在其他的路由器上查看。
3. RIP-1 是一种基于类别的路由协议。

八、配置序列

```
Router-A#sh run
Building configuration...
Current configuration : 1177 bytes
!
version 12.4
service timestamps debug datetime msec
service timestamps log datetime msec
no service password-encryption
!
hostname Router-A
!
boot-start-marker
boot-end-marker
!
enable password digitalchina
!
no aaa new-model
memory-size iomem 5
!
ip cef
ip ftp source-interface Serial1/1
no ip domain lookup
!
username cisco password 0 123456
!
interface FastEthernet0/0
 ip address 192.168.0.1 255.255.255.0
 duplex auto
 speed auto
!
interface Serial1/0
 no ip address
 shutdown
 serial restart-delay 0
 clock rate 64000
!
interface Serial1/1
```

```
    ip address 192.168.5.1 255.255.255.0
    serial restart-delay 0
    clock rate 64000
!
interface Serial1/2
    no ip address
    shutdown
    serial restart-delay 0
!
interface Serial1/3
    no ip address
    shutdown
    serial restart-delay 0
!
router ospf 1
    log-adjacency-changes
    redistribute connected subnets
    redistribute static subnets
    network 192.168.5.0 0.0.0.255 area 0
!
ip http server
no ip http secure-server
ip route 191.13.2.0 255.255.255.0 192.168.0.4
```

九、思考题

1. 在配置 RIP 和 OSPF 协议时，为什么不宣告 192.168.0.0？
2. 如果配置静态路由时的下一跳是 192.168.5.2，会导致什么结果？

十、课后练习

场景描述：假设你是一家中型企业的网络管理员，公司网络中使用 RIP 和 OSPF 两种路由协议。你负责管理子网 10.10.10.0/24，该子网需要通过特定的下一跳地址 192.168.0.9 访问。你需要将这条静态路由引入到 RIP 和 OSPF 路由域中，以确保整个网络的连通性。

1. 网络拓扑设计：设计一个包含至少三个路由器的网络拓扑，其中两个路由器运行 RIP 协议，另一个路由器运行 OSPF 协议。
2. 基本配置：为每个路由器配置接口 IP 地址，并在适当的路由器上配置 RIP 和 OSPF 协议。
3. 静态路由配置：在其一个路由器上配置静态路由 10.10.10.0/24，下一跳地址为 192.168.0.9。
4. 路由重分发：将上述静态路由引入到 RIP 和 OSPF 协议中，以确保整个网络可以访问 10.10.10.0/24 网络。
5. 测试与验证：从 RIP 和 OSPF 网络中的不同路由器尝试 ping 10.10.10.0/24 网络，以验证路由是否成功引入。

实验十 单区域 OSPF 基本配置

一、实验目的

1. 掌握单区域 OSPF 的配置。
2. 理解链路状态路由协议的工作过程。
3. 掌握实验环境中虚拟接口的配置。

二、相关知识

OSPF 是一种链路状态路由协议，它通过交换链路状态信息来构建网络拓扑图，并使用迪杰斯特拉算法计算到达每个网络的最短路径。

本实验配置适用的场景：OSPF 通常应用于中大型企业网络，因为它能够处理大量的路由信息和复杂的网络拓扑。

三、实验设备

1. DCR-1751 2 台。
2. CR-V35MT 1 条。
3. CR-V35FC 1 条。

四、实验拓扑

该实验拓扑结构如图 2-20 所示。

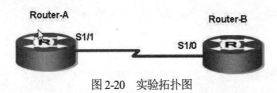

图 2-20 实验拓扑图

五、实验要求

实验中各设备的接口配置如表 2-9 所示。

表 2-9 设备接口配置

Router-A		Router-B	
S1/1	192.168.5.1/24	S1/0	192.168.5.2/24
Loopback0	10.10.10.1/24	Loopback0	10.10.11.1/24

六、实验步骤

第一步：路由器(Router-A 和 Router-B)环回接口的配置如下(其他接口的配置请参见实

验三)。

Router-A:

Router-A(config)#int loopback 0
Router-A(config-if)#ip address 10.10.10.1 255.255.255.0

Router-B:

Router-B#config terminal
Router-B(config)#int loopback 0
Router-B(config-if)#ip address 10.10.11.1 255.255.255.0

第二步：验证接口配置。

Router-B#sh interface loopback0
Loopback0 is up, line protocol is up
　Hardware is Loopback
　Internet address is 10.10.11.1/24
　MTU 1514 bytes, BW 8000000 Kbit, DLY 5000 usec,
　　reliability 255/255, txload 1/255, rxload 1/255
　Encapsulation LOOPBACK, loopback not set

第三步：路由器的 OSPF 配置。
Router-A 的配置：

Router-A(config)#router ospf 2　　　　　　　　　　　　！启动 OSPF 进程，进程号为 2
Router-A(config-router)#network 10.10.10.0 255.255.255.0 area 0　！注意要写掩码和区域号
Router-A(config-router)#network 192.168.5.0 255.255.255.0 area 0

Router-B 的配置：

Router-B(config)#router ospf 1
Router-B(config-router)#network 10.10.11.0 255.255.255.0 area 0
Router-B(config-router)#network 192.168.5.0 255.255.255.0 area 0

第四步：查看路由表。
Router-A:

Router-A#sh ip route
Codes: C - connected, S - static, R - RIP, M - mobile, B - BGP
　　　　D - EIGRP, EX - EIGRP external, O - OSPF, IA - OSPF inter area
　　　　N1 - OSPF NSSA external type 1, N2 - OSPF NSSA external type 2
　　　　E1 - OSPF external type 1, E2 - OSPF external type 2
　　　　i - IS-IS, su - IS-IS summary, L1 - IS-IS level-1, L2 - IS-IS level-2
　　　　ia - IS-IS inter area, * - candidate default, U - per-user static route

         o - ODR, P - periodic downloaded static route

Gateway of last resort is not set

C   192.168.5.0/24 is directly connected, Serial1/1
        10.0.0.0/8 is variably subnetted, 2 subnets, 2 masks
O   10.10.11.1/32 [110/65] via 192.168.5.2, 00:00:20, Serial1/1   ！注意到环回接口产生的是主机路由
C   10.10.10.0/24 is directly connected, Loopback0

Router-B:

Router-B#show ip route
Codes: C - connected, S - static, R - RIP, M - mobile, B - BGP
       D - EIGRP, EX - EIGRP external, O - OSPF, IA - OSPF inter area
       N1 - OSPF NSSA external type 1, N2 - OSPF NSSA external type 2
       E1 - OSPF external type 1, E2 - OSPF external type 2
       i - IS-IS, su - IS-IS summary, L1 - IS-IS level-1, L2 - IS-IS level-2
       ia - IS-IS inter area, * - candidate default, U - per-user static route
       o - ODR, P - periodic downloaded static route

Gateway of last resort is not set

C   192.168.5.0/24 is directly connected, Serial1/1
        10.0.0.0/8 is variably subnetted, 2 subnets, 2 masks
C   10.10.11.0/24 is directly connected, Loopback0
O   10.10.10.1/32 [110/65] via 192.168.5.1, 00:01:22, Serial1/1

第五步：其他验证命令。

Router-B#sh ip ospf 1   ！显示该 OSPF 进程的信息
Routing Process "ospf 1" with ID 10.10.11.1
  Start time: 17:23:19.948, Time elapsed: 00:02:50.848
  Supports only single TOS(TOS0) routes
  Supports opaque LSA
  Supports Link-local Signaling (LLS)
    Area BACKBONE(0)
        Number of interfaces in this area is 2 (1 loopback)
        Area has no authentication
        SPF algorithm last executed 00:02:30.720 ago
        SPF algorithm executed 2 times
        Area ranges are
        Number of LSA 2. Checksum Sum 0x01DC8C
        Number of opaque link LSA 0. Checksum Sum 0x000000
        Number of DCbitless LSA 0

		Number of indication LSA 0
		Number of DoNotAge LSA 0
		Flood list length 0
Router-A#show ip ospf interace   ！显示 OSPF 接口状态和类型
Serial1/1 is up, line protocol is up
    Internet Address 192.168.5.1/24, Area 0
    Process ID 2, Router ID 10.10.10.1, Network Type POINT_TO_POINT, Cost: 64
    Transmit Delay is 1 sec, State POINT_TO_POINT,
    Timer intervals configured, Hello 10, Dead 40, Wait 40, Retransmit 5
        oob-resync timeout 40
        Hello due in 00:00:06
    Supports Link-local Signaling (LLS)
    Index 2/2, flood queue length 0
    Next 0x0(0)/0x0(0)
    Last flood scan length is 1, maximum is 1
    Last flood scan time is 0 msec, maximum is 0 msec
    Neighbor Count is 1, Adjacent neighbor count is 1
    Adjacent with neighbor 10.10.11.1
    Suppress hello for 0 neighbor(s)
Loopback0 is up, line protocol is up
    Internet Address 10.10.10.1/24, Area 0
    Process ID 2, Router ID 10.10.10.1, Network Type LOOPBACK, Cost: 1
    Loopback interface is treated as a stub Host

Router-A#sh ip ospf neighbor   ！显示 OSPF 邻居

Neighbor ID	Pri	State	Dead Time	Address	Interface
10.10.11.1	0	FULL/ -	00:00:30	192.168.5.2	Serial1/1

第六步：修改环回接口的网络类型。

Router-A#config t
Router-A(config)#int loopback 0
Router-A(config-if)#ip ospf network point-to-point   ！将类型改为点到点

第七步：查看接口状态和 Router-B 的路由表。

Router-A#sh ip ospf interface
Loopback0 is up, line protocol is up
    Internet Address 10.10.10.1/24, Area 0
    Process ID 2, Router ID 10.10.10.1, Network Type POINT_TO_POINT, Cost: 1
    Transmit Delay is 1 sec, State POINT_TO_POINT,
    Timer intervals configured, Hello 10, Dead 40, Wait 40, Retransmit 5
        oob-resync timeout 40

    Supports Link-local Signaling (LLS)

    Index 1/1, flood queue length 0

    Next 0x0(0)/0x0(0)

    Last flood scan length is 0, maximum is 0

    Last flood scan time is 0 msec, maximum is 0 msec

    Neighbor Count is 0, Adjacent neighbor count is 0

    Suppress hello for 0 neighbor(s)

Serial1/1 is up, line protocol is up

    Internet Address 192.168.5.1/24, Area 0

    Process ID 2, Router ID 10.10.10.1, Network Type POINT_TO_POINT, Cost: 64

    Transmit Delay is 1 sec, State POINT_TO_POINT,

    Timer intervals configured, Hello 10, Dead 40, Wait 40, Retransmit 5

        oob-resync timeout 40

        Hello due in 00:00:07

    Supports Link-local Signaling (LLS)

    Index 2/2, flood queue length 0

    Next 0x0(0)/0x0(0)

    Last flood scan length is 1, maximum is 1

    Last flood scan time is 0 msec, maximum is 0 msec

    Neighbor Count is 1, Adjacent neighbor count is 1

        Adjacent with neighbor 10.10.11.1

    Suppress hello for 0 neighbor(s)

Router-B#sh ip route

Codes: C - connected, S - static, R - RIP, M - mobile, B - BGP
       D - EIGRP, EX - EIGRP external, O - OSPF, IA - OSPF inter area
       N1 - OSPF NSSA external type 1, N2 - OSPF NSSA external type 2
       E1 - OSPF external type 1, E2 - OSPF external type 2
       i - IS-IS, su - IS-IS summary, L1 - IS-IS level-1, L2 - IS-IS level-2
       ia - IS-IS inter area, * - candidate default, U - per-user static route
       o - ODR, P - periodic downloaded static route

Gateway of last resort is not set

C   192.168.5.0/24 is directly connected, Serial1/1
     10.0.0.0/24 is subnetted, 2 subnets
O   10.10.10.0 [110/65] via 192.168.5.1, 00:01:22, Serial1/1
C   10.10.11.0 is directly connected, Loopback0

七、注意事项和排错

1. 每个路由器的 OSPF 进程号可以不同，一个路由器可以有多个 OSPF 进程。
2. OSPF 是一种无类路由协议，因此在配置时必须指定子网掩码。
3. 第一个区域必须是区域 0。

八、配置序列

```
Router-A#show running-conf
Building configuration...
Current configuration : 1254 bytes
!
version 12.4
service timestamps debug datetime msec
service timestamps log datetime msec
no service password-encryption
!
hostname Router-A
!
boot-start-marker
boot-end-marker
!
enable password digitalchina
!
no aaa new-model
memory-size iomem 5
!
ip cef
ip ftp source-interface Serial1/1
no ip domain lookup
!
username cisco password 0 123456
!
interface Loopback0
  ip address 10.10.10.1 255.255.255.0
  ip ospf network point-to-point
!
interface FastEthernet0/0
  ip address 192.168.0.1 255.255.255.0
  shutdown
  duplex auto
  speed auto
!
interface Serial1/0
  no ip address
  shutdown
  serial restart-delay 0
  clock rate 64000
!
interface Serial1/1
```

```
    ip address 192.168.5.1 255.255.255.0
    serial restart-delay 0
    clock rate 64000
!
router ospf 2
    log-adjacency-changes
    network 10.10.10.0 0.0.0.255 area 0
    network 192.168.5.0 0.0.0.255 area 0
!
ip http server
no ip http secure-server
ip route 191.13.2.0 255.255.255.0 192.168.0.4
!
```

九、思考题

1. OSPF 与 RIP 的区别是什么？
2. 环回接口有什么好处？

十、课后练习

场景描述：假设你是一家快速发展的科技公司的网络管理员。公司近期搬迁到一个新的办公地点，要求你为新网络配置 OSPF 协议，以确保网络环境的稳定和通信效率。你需要配置一个单区域 OSPF 网络，实现公司内部所有部门能够互相通信。

1. 网络拓扑设计：设计一个单区域 OSPF 网络拓扑，至少包括 3 台路由器，每台路由器至少连接两台 PC。
2. 基本配置：为每台路由器的接口配置 IP 地址，并根据需要为交换机配置 VLAN。
3. OSPF 配置：在所有路由器上配置单区域 OSPF，包括设置 OSPF 网络类型，为每个路由器接口配置 OSPF 区域，并根据需要进行设置。
4. 路由 ID 配置：为每台路由器配置 Router ID。
5. 邻居关系建立：配置 OSPF 邻居关系，确保所有路由器之间能够建立邻居关系。
6. 测试和验证：通过从不同部门的 PC 进行 ping 测试，验证 OSPF 配置的正确性。

实验十一 RIP-2 邻居认证配置

一、实验目的

1. 掌握邻居认证的配置。
2. 理解 RIP-2 与 RIP-1 的区别。

二、相关知识

RIP-2(RIP 版本 2)支持一种简单的密码认证机制，允许网络管理员为 RIP 路由更新设置密码，从而增强网络的安全性。认证类型包括明文认证和密文认证。

RIP-2 邻居认证配置主要应用以下场景。

(1) 企业内部网络。在企业内部网络中，当需要防止未授权路由器接入网络时，使用 RIP-2 邻居认证可以提供额外的安全保障。

(2) 远程访问网络。在远程访问网络中，RIP-2 邻居认证可以限制对远程网络的访问，确保只有经过认证的路由器才能连接到远程网络。

三、实验设备

1. DCR-1751 2 台。
2. CR-V35MT 1 条。
3. CR-V35FC 1 条。

四、实验拓扑

该实验拓扑结构如图 2-21 所示。

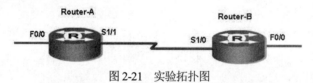

图 2-21　实验拓扑图

五、实验要求

实验中各设备的接口配置如表 2-10 所示。

表 2-10　设备接口配置

Router-A		Router-B	
S1/1(DCE)	192.168.5.1/24	S1/0(DTE)	192.168.5.2/24
F0/0	192.168.0.1	F0/0	192.169.2.1/24

六、实验步骤

第一步：参照实验三和表 2-10 配置路由器(Router-A 和 Router-B)的接口地址并测试网络连通性。

第二步：配置 Router-B。

```
Router-B#config t
Router-B(config)#router rip
Router-B(config-router)#version 2
Router-B(config-router)#network 192.168.5.0
```

```
Router-B(config-router)#network 192.168.2.0
Router-B(config-router)#exit
Router-B_config_rip#exit
Router-B(config)#int s1/0                              ! 进入与 Router-A 相连的接口
Router-B(config-if)#ip rip authentication mode text    ! 配置以明文方式验证
Router-B(config-if)#ip rip authentication key-chain cisco  ! 配置密码为 cisco
Router-B(config-if)#^Z
```

第三步：查看 Router-B 的配置。

```
Router-B#sh run
Building configuration...
Current configuration : 1022 bytes
!
version 12.4
service timestamps debug datetime msec
service timestamps log datetime msec
no service password-encryption
!
hostname Router-B
!
boot-start-marker
boot-end-marker
!
no aaa new-model
memory-size iomem 5
!
ip cef
no ip domain lookup
!
interface Loopback0
 no ip address
!
interface FastEthernet0/0
 ip address 192.168.2.1 255.255.255.0
 shutdown
 duplex auto
 speed auto
!
interface Serial1/0
 ip address 192.168.5.2 255.255.255.0
 ip rip authentication key-chain cisco
 serial restart-delay 0
!
router rip
 version 2
```

network 192.168.2.0
network 192.168.5.0

第四步：配置 Router-A(不启用认证)并查看其路由表。

Router-A(config)#router rip
Router-A(config-router)#version 2
Router-A(config-router)#network 192.168.0.0
Router-A(config-router)#network 192.168.5.0
Router-A(config-router)#^Z

查看路由表。

Router-A#sh ip route
Codes: C - connected, S - static, R - RIP, M - mobile, B - BGP
       D - EIGRP, EX - EIGRP external, O - OSPF, IA - OSPF inter area
       N1 - OSPF NSSA external type 1, N2 - OSPF NSSA external type 2
       E1 - OSPF external type 1, E2 - OSPF external type 2
       i - IS-IS, su - IS-IS summary, L1 - IS-IS level-1, L2 - IS-IS level-2
       ia - IS-IS inter area, * - candidate default, U - per-user static route
       o - ODR, P - periodic downloaded static route

Gateway of last resort is not set

C   192.168.5.0/24 is directly connected, Serial1/1   ！没有学习到路由
C   192.168.0.0/24 is directly connected, FastEthernet0/0

第五步：配置 Router-A 的认证设置。

Router-A(config)#key chain kal
Router-A(config-keychain)#key 1
Router-A(config-keychain-key)#key-string cisco
Router-A(config-keychain-key)#^Z
Router-A#config t
Router-A(config)#int s1/1
Router-A(config-if)#ip rip authentication key-chain kal
Router-A#sh ip route
Codes: C - connected, S - static, R - RIP, M - mobile, B - BGP
       D - EIGRP, EX - EIGRP external, O - OSPF, IA - OSPF inter area
       N1 - OSPF NSSA external type 1, N2 - OSPF NSSA external type 2
       E1 - OSPF external type 1, E2 - OSPF external type 2
       i - IS-IS, su - IS-IS summary, L1 - IS-IS level-1, L2 - IS-IS level-2
       ia - IS-IS inter area, * - candidate default, U - per-user static route
       o - ODR, P - periodic downloaded static route

Gateway of last resort is not set

C   192.168.5.0/24 is directly connected, Serial1/1
C   192.168.0.0/24 is directly connected, FastEthernet0/0
R   192.168.2.0/24 [120/1] via 192.168.5.2, 00:00:24, Serial1/1

```
Router-B#show ip route
Codes: C - connected, S - static, R - RIP, M - mobile, B - BGP
       D - EIGRP, EX - EIGRP external, O - OSPF, IA - OSPF inter area
       N1 - OSPF NSSA external type 1, N2 - OSPF NSSA external type 2
       E1 - OSPF external type 1, E2 - OSPF external type 2
       i - IS-IS, su - IS-IS summary, L1 - IS-IS level-1, L2 - IS-IS level-2
       ia - IS-IS inter area, * - candidate default, U - per-user static route
       o - ODR, P - periodic downloaded static route

Gateway of last resort is not set

C    192.168.5.0/24 is directly connected, Serial1/0
R    192.168.0.0/24 [120/1] via 192.168.5.1, 00:01:39, Serial1/0
C    192.168.2.0/24 is directly connected, FastEthernet0/0
```

七、注意事项和排错

1. 只有 RIP-2 支持认证功能。
2. 认证配置应在相邻的接口上进行。
3. 认证密码必须一致，并且双方都需进行配置。

八、配置序列

```
Router-B#sh run
Building configuration...
Current configuration : 1019 bytes
!
version 12.4
service timestamps debug datetime msec
service timestamps log datetime msec
no service password-encryption
!
hostname Router-B
!
boot-start-marker
boot-end-marker
!
no aaa new-model
memory-size iomem 5
!
ip cef
no ip domain lookup
!
key chain ka1
 key 1
  key-string cisco
!
```

```
interface FastEthernet0/0
 ip address 192.168.2.1 255.255.255.0
 duplex auto
 speed auto
!
interface Serial1/0
 ip address 192.168.5.2 255.255.255.0
 ip rip authentication key-chain ka1
 serial restart-delay 0
!
interface Serial1/1
 no ip address
 shutdown
 serial restart-delay 0
!
interface Serial1/2
 no ip address
 shutdown
 serial restart-delay 0
!
interface Serial1/3
 no ip address
 shutdown
 serial restart-delay 0
!
router rip
 version 2
 network 192.168.2.0
 network 192.168.5.0
 no auto-summary
!
```

九、思考题

1. RIP-2 认证有什么意义？
2. 为什么 RIP-2 认证一定要是双向的？

实验十二 多区域 OSPF 配置

一、实验目的

1. 掌握开放最短路径优先(OSPF)协议的多区域配置及其运作方式。

2. 能够在路由器上配置多区域 OSPF。
3. 通过多区域 OSPF 实现不同区域之间的互联，确保网络层的连通性。

二、相关知识

OSPF 是一种链路状态路由协议，通过交换链路状态信息来构建网络拓扑图。OSPF 网络被划分为多个区域，每个区域包含一组路由器，这些路由器通过 OSPF 协议交换路由信息。

在大规模网络中，为了减少资源消耗并将拓扑变化限制在局部区域，通常会进行区域划分。多区域 OSFP 配置实验主要用于以下场景。

(1) 中大型企业网络。多区域 OSPF 适用于中大型企业网络，因为它能够处理大量的路由信息和复杂的网络拓扑。

(2) 多区域网络设计。在需要划分多个 OSPF 区域以优化路由信息交换并减少路由器资源消耗的网络环境中，多区域 OSPF 特别有用。

三、实验设备

1. DCR-1751 3 台。
2. CR-V35MT 1 条。
3. CR-V35FC 1 条。

四、实验拓扑

该实验拓扑结构如图 2-22 所示。

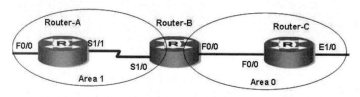

图 2-22 实验拓扑结构

五、实验要求

实验中各设备的接口配置如表 2-11 所示。

表 2-11 设备接口配置

Router-A		Router-B		Router-C	
S1/1(DCE)	192.168.5.1	S1/0(DTE)	192.168.5.2	F0/0	192.168.2.2
F0/0	192.168.0.1	F0/0	192.168.2.1	E1/0	192.168.3.1

其中 Router-B 为 ABR。

六、实验步骤

第一步：参照表 2-11 配置路由器的所有接口地址并测试网络连通性。
第二步：配置 Router-A。

```
Router-A#config t
Router-A(config)#router ospf 100
Router-A(config-router)#network 192.168.0.0 255.255.255.0 area 1
Router-A(config-router)#network 192.168.5.0 255.255.255.0 area 1
Router-A(config-router)#^Z
```

第三步：配置 Router-B。

```
Router-B#config t
Router-B(config)#router ospf 100
Router-B(config-router)#network 192.168.5.0 255.255.255.0 area 1
Router-B(config-router)#network 192.168.2.0 255.255.255.0 area 0
Router-B(config-router)#^Z
```

第四步：配置 Router-C。

```
Router-C#config t
Router-C(config)#router ospf 100
Router-C(config-router)#network 192.168.2.0 255.255.255.0 area 0
Router-C(config-router)#network 192.168.3.0 255.255.255.0 area 0
Router-C(config-router)#^Z
```

第五步：查看路由表。

```
Router-A#sh ip route
Codes: C - connected, S - static, R - RIP, M - mobile, B - BGP
       D - EIGRP, EX - EIGRP external, O - OSPF, IA - OSPF inter area
       N1 - OSPF NSSA external type 1, N2 - OSPF NSSA external type 2
       E1 - OSPF external type 1, E2 - OSPF external type 2
       i - IS-IS, su - IS-IS summary, L1 - IS-IS level-1, L2 - IS-IS level-2
       ia - IS-IS inter area, * - candidate default, U - per-user static route
       o - ODR, P - periodic downloaded static route

Gateway of last resort is not set

C    192.168.5.0/24 is directly connected, Serial1/1
C    192.168.0.0/24 is directly connected, FastEthernet0/0
O IA 192.168.2.0/24 [110/65] via 192.168.5.2, 00:02:50, Serial1/1
O IA 192.168.3.0/24 [110/75] via 192.168.5.2, 00:01:50, Serial1/1    ！区域间的路由
```

```
Router-B#sh ip route
Codes: C - connected, S - static, R - RIP, M - mobile, B - BGP
       D - EIGRP, EX - EIGRP external, O - OSPF, IA - OSPF inter area
       N1 - OSPF NSSA external type 1, N2 - OSPF NSSA external type 2
       E1 - OSPF external type 1, E2 - OSPF external type 2
       i - IS-IS, su - IS-IS summary, L1 - IS-IS level-1, L2 - IS-IS level-2
       ia - IS-IS inter area, * - candidate default, U - per-user static route
       o - ODR, P - periodic downloaded static route

Gateway of last resort is not set

C    192.168.5.0/24 is directly connected, Serial1/0
O    192.168.0.0/24 [110/65] via 192.168.5.1, 00:03:16, Serial1/0
C    192.168.2.0/24 is directly connected, FastEthernet0/0
O    192.168.3.0/24 [110/11] via 192.168.2.2, 00:02:16, FastEthernet0/0
                 ! 对 ABR 来说是区域内的路由
Router-C#sh ip route
Codes: C - connected, S - static, R - RIP, M - mobile, B - BGP
       D - EIGRP, EX - EIGRP external, O - OSPF, IA - OSPF inter area
       N1 - OSPF NSSA external type 1, N2 - OSPF NSSA external type 2
       E1 - OSPF external type 1, E2 - OSPF external type 2
       i - IS-IS, su - IS-IS summary, L1 - IS-IS level-1, L2 - IS-IS level-2
       ia - IS-IS inter area, * - candidate default, U - per-user static route
       o - ODR, P - periodic downloaded static route

Gateway of last resort is not set

O IA 192.168.5.0/24 [110/65] via 192.168.2.1, 00:02:58, FastEthernet0/0
O IA 192.168.0.0/24 [110/66] via 192.168.2.1, 00:02:58, FastEthernet0/0
C    192.168.2.0/24 is directly connected, FastEthernet0/0
C    192.168.3.0/24 is directly connected, Ethernet1/0
```

七、注意事项和排错

1. 区域划分应在接口上进行。
2. 必须有 area 0 存在。

八、配置序列

```
Router-B#sh run
Building configuration...

Current configuration : 1016 bytes
!
version 12.4
service timestamps debug datetime msec
```

```
service timestamps log datetime msec
no service password-encryption
!
hostname Router-B
!
boot-start-marker
boot-end-marker
!
no aaa new-model
memory-size iomem 5
!
ip cef
no ip domain lookup
!
interface FastEthernet0/0
 ip address 192.168.2.1 255.255.255.0
 duplex auto
 speed auto
!
interface Serial1/0
 ip address 192.168.5.2 255.255.255.0
 serial restart-delay 0
!
router ospf 100
 log-adjacency-changes
 network 192.168.2.0 0.0.0.255 area 0
 network 192.168.5.0 0.0.0.255 area 1
!
```

九、思考题

1. 为什么必须有 area 0 存在？
2. 在 Router-A 和 Router-C 声明网段时，是否有其他的方法可以使用？

实训项目十一　多路由器间的 OSPF 路由配置

一、实训目的

1. 理解 OSPF(开放最短路径优先)协议的工作原理。
2. 学会在多路由器环境中配置 OSPF。
3. 通过 OSPF 协议实现复杂网络中不同区域和 VLAN 之间的互联。

二、实训设备

(1) 二层交换机(S-A 和 S-B) 2 台。
(2) 路由器(R-C 和 R-D) 2 台。
(3) PC 4 台。
(4) 网线 7 根。

三、实训拓扑

该实验拓扑结构如图 2-23 所示。

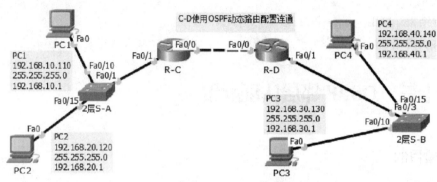

图 2-23　实验拓扑结构

四、实训要求

按照图 2-23 所示进行设备连线，并对各设备进行以下配置。
(1) 交换机 A 划分 vl 10(9~12)、vl 20(13~16)，Trunk (1)。
(2) 交换机 B 划分 vl 30(17~20)、vl 40(21~24)，Trunk (3)。
(3) 设置路由器 C：

E 0/1.10	192.168.10.1/24	封装 vl 10;
E 0/1.20	192.168.20.1/24	封装 vl 20;
F 0/0	192.168.100.1/30	

(4) 设置路由器 D：

F 0/1.30	192.168.30.1/24	封装 vl 30;
F 0/1.40	192.168.40.1/24	封装 vl 40;
F 0/0	192.168.100.2/30	

(5) 启用路由器的动态 OSPF 协议，配置路由器 C 的动态路由 OSPF 协议：

RC(config)#router ospf 1
RC(config-router)#network 192.168.10.0 255.255.255.0 area 0
RC(config-router)#network 192.168.20.0 255.255.255.0 area 0
RC(config-router)#network 192.168.100.0 255.255.255.252 area 0

(6) 配置路由器 D 的动态路由 OSPF 协议：

RD(config)#router ospf　2
RD(config-router)#network 192.168.30.0 255.255.255.0 area 0
RD(config-router)#network 192.168.40.0 255.255.255.0 area 0
RD(config-router)#network 192.168.100.0 255.255.255.252 area 0

验证：sh ip ospf　　　　sh ip route

(7) 根据连线位置正确配置 PC1 和 PC2 的地址(注意配置网关地址，并确保与 VLAN 的对应关系)。

实验结果：PC1---ping---PC2、3、4　　通。

查看设备状态：show vlan　　　show run　　　show ip route。

实验十三　OSPF 邻居认证配置

一、实验目的

1. 理解 OSPF 协议中的认证机制。
2. 路由器上配置 OSPF 邻居认证。
3. 配置 OSPF 邻居认证，以确保网络中不同区域和 VLAN 之间的互联。

二、相关知识

OSPF 邻居认证可以提高网络的安全性，防止未授权的路由器接入网络。在企业环境中，需要配置认证以确保 OSPF 路由的安全性。OSPF 认证类型包括明文认证和密文认证。OSPF 邻居认证主要应用于以下场景。

(1) 企业内部网络。在企业内部网络中，当需要防止未授权的路由器接入网络时，使用 OSPF 邻居认证可以提供额外的安全层。

(2) 远程访问网络。在远程访问网络中，OSPF 邻居认证可以用来限制对远程网络的访问，确保只有经过认证的路由器才能访问远程网络。

三、实验设备

1. DCR-1751 2 台。
2. CR-V35MT 1 条。
3. CR-V35FC 1 条。

四、实验拓扑

该实验拓扑结构如图 2-24 所示。

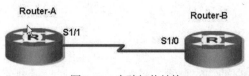

图 2-24 实验拓扑结构

五、实验要求

实验中各设备的接口配置如表 2-12 所示。

表 2-12 设备接口配置

Router-A		Router-B	
S1/1	192.168.5.1/24	S1/0	192.168.5.2/24
Loopback0	10.10.10.1/24	Loopback0	10.10.11.1/24

六、实验步骤

第一步： 路由器(Router-A 和 Router-B)环回接口的配置如下(其他接口配置可参见实验三)。
配置 Router-A：

```
Router-A(config)#interface   loopback0
Router-A(config-if)#ip address 10.10.10.1 255.255.255.0
```

配置 Router-B：

```
Router-B#config
Router-B(config)#interface loopback0
Router-B(config-if)#ip address   10.10.11.1 255.255.255.0
```

第二步： 验证接口配置。

```
Router-B#sh interface loopback0
Loopback0 is up, line protocol is up
    Hardware is Loopback
    Internet address is 10.10.11.1/24
    MTU 1514 bytes, BW 8000000 Kbit, DLY 5000 usec,
       reliability 255/255, txload 1/255, rxload 1/255
    Encapsulation LOOPBACK, loopback not set
```

第三步： 路由器的 OSPF 配置。
配置 Router-A：

```
Router-A(config)#router ospf 2                                !启动 OSPF 进程，进程号为 2
Router-A(config-router)#network 10.10.10.0 255.255.255.0 area 0   !注意要写掩码和区域号
Router-A(config-router)#network 192.168.5.0 255.255.255.0 area 0
```

Router-A(config-router)#area 0 authentication	！定义在区域 0 中使用明文认证
Router-A(config-router)#exit	
Router-A(config)#int s1/1	
Router-A(config-if)#ip ospf authentication-key cisco	！配置接口密码

配置 Router-B：

Router-B(config)#router ospf 1	
Router-B(config-router)#network 10.10.11.0 255.255.255.0 area 0	
Router-B(config-router)#network 192.168.5.0 255.255.255.0 area 0	
Router-B(config-router)#area 0 authentication	！定义在区域 0 中使用明文认证
Router-B(config-router)#exit	
Router-B(config)#int s1/0	
Router-B(config-if)#ip ospf authentication-key cisco	！配置接口密码

第四步：查看路由表。

Router-A：

```
Router-A#sh ip route
Codes: C - connected, S - static, R - RIP, M - mobile, B - BGP
       D - EIGRP, EX - EIGRP external, O - OSPF, IA - OSPF inter area
       N1 - OSPF NSSA external type 1, N2 - OSPF NSSA external type 2
       E1 - OSPF external type 1, E2 - OSPF external type 2
       i - IS-IS, su - IS-IS summary, L1 - IS-IS level-1, L2 - IS-IS level-2
       ia - IS-IS inter area, * - candidate default, U - per-user static route
       o - ODR, P - periodic downloaded static route

Gateway of last resort is not set

C    192.168.5.0/24 is directly connected, Serial1/1
     10.0.0.0/8 is variably subnetted, 2 subnets, 2 masks
O       10.10.11.1/32 [110/65] via 192.168.5.2, 00:00:46, Serial1/1
C       10.10.10.0/24 is directly connected, Loopback0
```

Router-B：

```
Router-B#show ip route
Codes: C - connected, S - static, R - RIP, M - mobile, B - BGP
       D - EIGRP, EX - EIGRP external, O - OSPF, IA - OSPF inter area
       N1 - OSPF NSSA external type 1, N2 - OSPF NSSA external type 2
       E1 - OSPF external type 1, E2 - OSPF external type 2
       i - IS-IS, su - IS-IS summary, L1 - IS-IS level-1, L2 - IS-IS level-2
       ia - IS-IS inter area, * - candidate default, U - per-user static route
       o - ODR, P - periodic downloaded static route
```

```
Gateway of last resort is not set
C    192.168.5.0/24 is directly connected, Serial1/0
     10.0.0.0/8 is variably subnetted, 2 subnets, 2 masks
C    10.10.11.0/24 is directly connected, Loopback0
O    10.10.10.1/32 [110/65] via 192.168.5.1, 00:00:15, Serial1/0
```

七、注意事项和排错

1. 在邻居接口上配置认证。
2. 认证方式除了明文认证，还有密钥认证。

八、配置序列

无。

九、思考题

1. 认证的作用是什么？
2. 认证应该在哪里配置？

十、课后练习

场景描述：假设你是一家金融机构的网络管理员。由于金融数据的安全性至关重要，公司要求你在内部的 OSPF 网络中实施邻居认证，以确保只有经过认证的路由器才能交换路由信息。

1. 网络拓扑设计：设计一个至少包含两个路由器的网络拓扑，每个路由器至少有一个接口连接到模拟的 WAN。确定每个路由器的接口 IP 地址和 OSPF 进程 ID。
2. 基本配置：为每个路由器的接口配置 IP 地址，并配置 OSPF 的基本设置。
3. OSPF 邻居认证配置：分别在两个路由器上配置 OSPF 邻居认证，选择认证类型并配置认证密钥。
4. 测试和验证：观察并记录路由器之间的 OSPF 邻居状态，并尝试修改认证密钥，以观察邻居关系的变化。

实验十四　OSPF 路由汇总配置

一、实验目的

1. 理解 OSPF 协议中的路由汇总机制。
2. 在路由器上配置 OSPF 路由汇总。
3. 通过配置 OSPF 路由汇总，优化网络性能，减少路由表，提高路由效率。

二、相关知识

路由汇总是指将多个网络的路由信息合并为一个较大的网络，以减少路由表中的路由条目数量。在大规模网络中，路由表往往非常庞大，从而降低了转发速度。通常在子网边界进行汇总可以减少路由表的长度。OSPF 的路由汇总包括区域间路由汇总和区域内路由汇总两种类型。

OSPF 路由汇总配置实验主要应用于以下场景。

（1）中大型企业网络。在中大型企业网络中，OSPF 路由汇总适用于需要管理大量路由信息的网络环境。

（2）多区域网络设计。在多区域网络中，OSPF 路由汇总可以用于优化路由信息的交换，减少路由器资源的消耗。

三、实验设备

1. DCR-1751 3 台。
2. CR-V35MT 1 条。
3. CR-V35FC 1 条。

四、实验拓扑

该实验拓扑结构如图 2-25 所示。

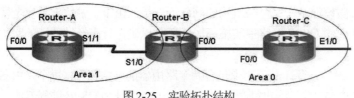

图 2-25　实验拓扑结构

五、实验要求

实验中各设备的接口配置如表 2-13 所示。

表 2-13　设备接口配置

Router-A		Router-B		Router-C	
S1/1(DCE)	10.10.11.1/24	S1/0(DTE)	10.10.11.2/24	F0/0	192.168.2.2/24
F0/0	10.10.10.1/24	F0/0	192.168.2.1/24	E1/0	192.168.3.1/24

其中 Router-B 为 ABR。

六、实验步骤

第一步：参照表 2-13 配置各接口地址，并测试网络连通性。
第二步：Router-A 的配置。

Router-A#config t

Router-A(config)#router ospf 100

Router-A(config-router)#network 10.10.10.0 255.255.255.0 area 1

Router-A(config-router)#network 10.10.11.0 255.255.255.0 area 1

Router-A(config-router)#^Z

第三步：Router-B 的配置。

Router-B#config t

Router-B(config)#router ospf 100

Router-B(config-router)#network 10.10.11.0 255.255.255.0 area 1   ！注意区域的划分在接口上

Router-B(config-router)#network 192.168.2.0 255.255.255.0 area 0

Router-B(config-router)#^Z

第四步：Router-C 的配置。

Router-C#config t

Router-C(config)#router ospf 100

Router-C(config-router)#network 192.168.2.0 255.255.255.0 area 0

Router-C(config-router)#network 192.168.3.0 255.255.255.0 area 0

Router-C(config-router)#^Z

第五步：查看路由表。

Router-A#sh ip route

Codes: C - connected, S - static, R - RIP, M - mobile, B - BGP

       D - EIGRP, EX - EIGRP external, O - OSPF, IA - OSPF inter area

       N1 - OSPF NSSA external type 1, N2 - OSPF NSSA external type 2

       E1 - OSPF external type 1, E2 - OSPF external type 2

       i - IS-IS, su - IS-IS summary, L1 - IS-IS level-1, L2 - IS-IS level-2

       ia - IS-IS inter area, * - candidate default, U - per-user static route

       o - ODR, P - periodic downloaded static route

Gateway of last resort is not set

    10.0.0.0/24 is subnetted, 2 subnets

C   10.10.10.0 is directly connected, FastEthernet0/0

C   10.10.11.0 is directly connected, Serial1/1

O IA 192.168.2.0/24 [110/65] via 10.10.11.2, 00:02:11, Serial1/1

O IA 192.168.3.0/24 [110/75] via 10.10.11.2, 00:00:31, Serial1/1 ！区域间的路由

Router-B#sh ip route

Codes: C - connected, S - static, R - RIP, M - mobile, B - BGP

       D - EIGRP, EX - EIGRP external, O - OSPF, IA - OSPF inter area

       N1 - OSPF NSSA external type 1, N2 - OSPF NSSA external type 2

         E1 - OSPF external type 1, E2 - OSPF external type 2
         i - IS-IS, su - IS-IS summary, L1 - IS-IS level-1, L2 - IS-IS level-2
         ia - IS-IS inter area, * - candidate default, U - per-user static route
         o - ODR, P - periodic downloaded static route

Gateway of last resort is not set

     10.0.0.0/24 is subnetted, 2 subnets
O   10.10.10.0 [110/65] via 10.10.11.1, 00:03:07, Serial1/0
C   10.10.11.0 is directly connected, Serial1/0
C   192.168.2.0/24 is directly connected, FastEthernet0/0
O   192.168.3.0/24 [110/11] via 192.168.2.2, 00:01:27, FastEthernet0/0
              ！对 ABR 来说是区域内的路由

Router-C#sh ip route
Codes: C - connected, S - static, R - RIP, M - mobile, B - BGP
       D - EIGRP, EX - EIGRP external, O - OSPF, IA - OSPF inter area
       N1 - OSPF NSSA external type 1, N2 - OSPF NSSA external type 2
       E1 - OSPF external type 1, E2 - OSPF external type 2
       i - IS-IS, su - IS-IS summary, L1 - IS-IS level-1, L2 - IS-IS level-2
       ia - IS-IS inter area, * - candidate default, U - per-user static route
       o - ODR, P - periodic downloaded static route

Gateway of last resort is not set

     10.0.0.0/24 is subnetted, 2 subnets
O IA   10.10.10.0 [110/66] via 192.168.2.1, 00:02:05, FastEthernet0/0   ！区域间的路由
O IA   10.10.11.0 [110/65] via 192.168.2.1, 00:02:05, FastEthernet0/0
C   192.168.2.0/24 is directly connected, FastEthernet0/0
C   192.168.3.0/24 is directly connected, Ethernet1/0

第六步：在 Router-B 上配置路由汇总。

Router-B#config t
Router-B(config)#router ospf 100
Router-B(config-router)#area 1 range 10.10.0.0 255.255.0.0
Router-B(config-router)#network 192.168.2.0 255.255.255.0 area 0
Router-B(config-router)#^Z

重新查看 Router-C 上的路由表。

Router-C#sh ip route
Codes: C - connected, S - static, R - RIP, M - mobile, B - BGP
       D - EIGRP, EX - EIGRP external, O - OSPF, IA - OSPF inter area
       N1 - OSPF NSSA external type 1, N2 - OSPF NSSA external type 2
       E1 - OSPF external type 1, E2 - OSPF external type 2

```
                i - IS-IS, su - IS-IS summary, L1 - IS-IS level-1, L2 - IS-IS level-2
                ia - IS-IS inter area, * - candidate default, U - per-user static route
                o - ODR, P - periodic downloaded static route

Gateway of last resort is not set

        10.0.0.0/16 is subnetted, 1 subnets
O IA    10.10.0.0 [110/65] via 192.168.2.1, 00:01:04, FastEthernet0/0   ！注意新的掩码
C       192.168.2.0/24 is directly connected, FastEthernet0/0
C       192.168.3.0/24 is directly connected, Ethernet1/0
```

七、注意事项和排错

1. 在实际环境中，通常会进行精确的路由汇总。
2. 路由汇总操作通常在边界路由器上执行。

八、配置序列

```
Router-B#sh run
Building configuration...

Current configuration : 1008 bytes
!
version 12.4
service timestamps debug datetime msec
service timestamps log datetime msec
no service password-encryption
!
hostname Router-B
!
boot-start-marker
boot-end-marker
!
no aaa new-model
memory-size iomem 5
!
ip cef
no ip domain lookup
!
interface FastEthernet0/0
 ip address 192.168.2.1 255.255.255.0
 duplex auto
 speed auto
```

```
!
interface Serial1/0
  ip address 10.10.11.2 255.255.255.0
  serial restart-delay 0
!
router ospf 100
  log-adjacency-changes
  area 1 range 10.10.0.0 255.255.0.0
  network 10.10.11.0 0.0.0.255 area 1
  network 192.168.2.0 0.0.0.255 area 0
!
```

九、思考题

1. 路由汇总的作用是什么？
2. 汇总操作通常在什么地方进行？

十、课后练习

场景描述：假设你是一家小型企业的网络管理员。公司分配给你一个地址块 10.0.0.0/25，要求你为公司的内部网络设计一个可扩展的 OSPF 网络，并进行路由汇总，以优化网络性能。

1. 网络拓扑设计：设计一个至少包含 3 个路由器的网络拓扑，使用 10.0.0.0/25 地址块进行子网划分。
2. 基本配置：为每个路由器的接口配置 IP 地址，使用规划好的子网地址，并进行 OSPF 的基本配置。
3. OSPF 路由汇总配置：在一个主路由器上配置 OSPF 路由汇总，以减少路由表中的条目。
4. 测试和验证：检查每个路由器的路由表，确认汇总路由是否存在；从不同子网的 PC 尝试 ping 其他子网的 PC，从而验证网络连通性。

实验十五 NAT 地址转换配置

一、实验目的

1. 掌握地址转换的配置方法。
2. 掌握将内部服务器地址发布到外部的方法。
3. 掌握私有地址访问互联网的配置方法。

二、相关知识

NAT 是一种将私有 IP 地址转换为公共 IP 地址的技术,它允许内部网络使用非注册的 IP 地址与外部网络进行通信。

NAT 类型包括以下几种。
- 静态 NAT:为内部网络中的每个私有 IP 地址分配一个固定的公共 IP 地址。
- 动态 NAT:为内部网络中的私有 IP 地址分配一个公共 IP 地址池,地址映射是动态的。
- PAT(端口地址转换):通过端口号来区分不同的内部地址,通常与动态 NAT 结合使用。

NAT 地址转换主要应用于以下场景。
(1) 企业内部有对互联网提供服务的 Web 服务器。
(2) 企业内部使用私有地址的主机需要访问互联网。

三、实验设备

1. DCR-1751 2 台。
2. PC 2 台。

四、实验拓扑

该实验拓扑结构如图 2-26 所示。

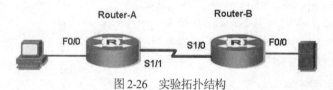

图 2-26　实验拓扑结构

五、实验要求

实验中各设备的接口配置如表 2-14 所示。

表 2-14　设备接口配置

Router-A		Router-B		PC		Server	
S1/1(DCE)	192.168.5.1/24	S1/0	192.168.5.2/24	IP	192.168.0.3/24	IP	192.168.2.2/24
F0/0	192.168.0.1/24	F0/0	192.168.2.1/24	网关	192.168.0.1	网关	192.168.2.1

六、实验步骤

内部 PC 需要访问外部的服务器。

假设在 Router-A 上做地址转换,将 192.168.0.0/24 转换成 192.168.5.10~192.168.5.20 之间的地址,并且实现端口的地址复用。

第一步:参照表 2-14 将接口地址和 PC 地址配置好,并测试网络连通性。

第二步:配置 Router-A 的 NAT。

```
Router-A#config t
Router-A(config)#ip access-list standard 1              !定义访问控制列表
Router-A(config-std-nacl)#permit 192.168.0.0   0.0.0.255    !定义允许转换的源地址范围
Router-A(config-std-nacl)#exit
Router-A(config)#ip nat pool overld 192.168.5.10 192.168.5.20 netmask 255.255.255.0
!定义名为 overld 的转换地址池
Router-A(config)#ip nat inside source list 1 pool overld overload
!配置将 ACL 允许的源地址转换成 overld 中的地址,并且做 PAT 的地址复用
Router-A(config)#int f0/0
Router-A(config-if)#ip nat inside                       !定义 F0/0 为内部接口
Router-A(config-if)#int s1/1
Router-A(config-if)#ip nat outside                      !定义 S1/1 为外部接口
Router-A(config-if)#exit
Router-A(config)#ip route 0.0.0.0 0.0.0.0 192.168.5.2   !配置路由器 A 的默认路由
```

第三步:查看 Router-B 的路由表。

```
Router-B#sh ip route
Codes: C - connected, S - static, I - IGRP, R - RIP, M - mobile, B - BGP
    D - EIGRP, EX - EIGRP external, O - OSPF, IA - OSPF inter area
    N1 - OSPF NSSA external type 1, N2 - OSPF NSSA external type 2
    E1 - OSPF external type 1, E2 - OSPF external type 2, E - EGP
    i - IS-IS, L1 - IS-IS level-1, L2 - IS-IS level-2, ia - IS-IS inter area
    * - candidate default, U - per-user static route, o - ODR
    P - periodic downloaded static route

Gateway of last resort is not set

C 192.168.2.0/24 is directly connected, FastEthernet0/0
C 192.168.5.0/24 is directly connected, Serial1/0
!注意:并没有到 192.168.0.0 的路由
```

第四步:测试。

测试结果如图 2-27 所示。

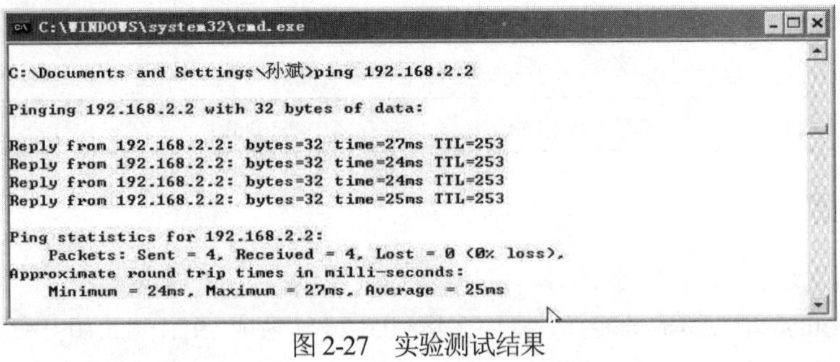

```
C:\WINDOWS\system32\cmd.exe

C:\Documents and Settings\孙斌>ping 192.168.2.2

Pinging 192.168.2.2 with 32 bytes of data:

Reply from 192.168.2.2: bytes=32 time=27ms TTL=253
Reply from 192.168.2.2: bytes=32 time=24ms TTL=253
Reply from 192.168.2.2: bytes=32 time=24ms TTL=253
Reply from 192.168.2.2: bytes=32 time=25ms TTL=253

Ping statistics for 192.168.2.2:
    Packets: Sent = 4, Received = 4, Lost = 0 (0% loss),
Approximate round trip times in milli-seconds:
    Minimum = 24ms, Maximum = 27ms, Average = 25ms
```

图 2-27　实验测试结果

第五步:查看地址转换表。

```
Router-A#sh ip nat translations
```

Pro Inside global Inside local Outside local Outside global
icmp 192.168.5.10:30 192.168.0.3:30 192.168.5.2:30 192.168.5.2:30
icmp 192.168.5.10:31 192.168.0.3:31 192.168.5.2:31 192.168.5.2:31
icmp 192.168.5.10:32 192.168.0.3:32 192.168.5.2:32 192.168.5.2:32
icmp 192.168.5.10:33 192.168.0.3:33 192.168.5.2:33 192.168.5.2:33

七、注意事项和排错

1. 注意转换的方向和接口。
2. 注意地址池和 ACL 的名称。
3. 需要配置 Router-A 的默认路由。

八、配置序列

```
Router-A#sh run
Building configuration...

Current configuration : 1115 bytes
!
version 12.4
no service timestamps log datetime msec
no service timestamps debug datetime msec
no service password-encryption
!
hostname Router-A
!
no ip cef
no ipv6 cef
!
spanning-tree mode pvst
!
interface FastEthernet0/0
ip address 192.168.0.1 255.255.255.0
ip nat inside
duplex auto
speed auto
!
interface FastEthernet0/1
no ip address
duplex auto
speed auto
shutdown
!
interface Serial1/0
```

```
no ip address
clock rate 2000000
shutdown
!
interface Serial1/1
ip address 192.168.5.1 255.255.255.0
ip nat outside
clock rate 64000
!
ip nat pool overld 192.168.5.10 192.168.5.20 netmask 255.255.255.0
ip nat inside source list 1 pool overld overload
ip classless
ip route 0.0.0.0 0.0.0.0 192.168.5.2
!
ip flow-export version 9
!
!
access-list 1 permit 0.0.0.0 255.255.255.0
access-list 1 permit 192.168.0.0 0.0.0.255
!
```

九、思考题

1. 为什么在 Router-B 上无需配置 192.168.0.0 的路由就能实现通信？
2. 为什么在 Router-A 上需要配置默认路由？
3. 如果外部接口地址是通过拨号动态获取的，该如何进行配置？
4. 请指出上述配置中的 Inside local、Inside global、Outside local、Outside global 分别是什么？

十、课后练习

场景描述：假设你是一家小型企业的网络管理员。公司内部有一台运行 FTP 服务的服务器，其 IP 地址为 192.168.2.2/24。你需要将这台服务器发布到外部网络，使外部用户能够访问其 FTP 服务。外部网络的地址为 192.168.5.2/24。你需要在 Router-B 上进行相应的 NAT 配置。

  1. 网络拓扑设计：设计网络拓扑，包括 Router-B、内部服务器、内部 PC 以及外部网络连接，确定内部和外部网络的接口 IP 地址。

  2. 服务器配置：在 IP 地址为 192.168.2.2/24 的服务器上安装并配置 FTP 服务，设置 FTP 服务器的用户账户和权限。

  3. 路由器配置：在 Router-B 配置内部接口和外部接口的 IP 地址，配置静态 NAT，将内部服务器的 IP 地址 192.168.2.2 映射到外部地址 192.168.5.2。

  4. 测试与验证：使用内部 PC 尝试访问外部地址 192.168.5.2 的 FTP 服务，验证 NAT 配置是否成功。

第 3 章

网络综合实践

实践一 二层接入三层连通加聚合实践项目

一、实践目的

1. 实现二层交换机与三层交换机之间的连通。
2. 配置 VLAN，实现不同网络段的隔离和聚合。
3. 诊断并排除网络配置过程中出现的网络故障。

二、实践拓扑

该实验拓扑结构如图 3-1 所示。

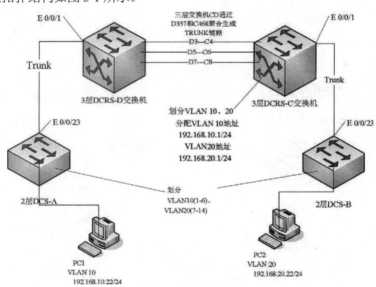

图 3-1 实验拓扑结构

三、实践步骤

1. 设备准备：连接二层交换机、三层交换机和 PC，并确保所有连接正确无误。
2. 网络规划：根据实验需求规划网络拓扑，确定交换机和 PC 的连接方式。

3. 设备配置：配置二层交换机的接口和 VLAN，配置三层交换机的接口、VLAN 及聚合链路。
4. 测试和验证：使用 ping 命令测试网络连通性，并验证配置是否正确。
5. 故障排除：进行网络故障诊断，解决存在的问题。
6. 文档编写：记录网络配置、更变情况和测试结果。

具体要求：如图 3-1 所示，首先在交换机 C 和交换机 D 上先启动生成树协议，并进行相应配置，使 PC1 和 PC2 之间能 ping 通。

实践二　单臂路由加聚合实践项目

一、实践目的

1. 使用单臂路由在路由器的一个物理接口上配置多个逻辑接口。
2. 配置聚合链路。
3. 通过单臂路由和聚合链路的配置，实现网络中不同子网之间的连通。
4. 诊断并排除网络配置过程中出现的网络故障。

二、实践拓扑

该实验拓扑结构如图 3-2 所示。

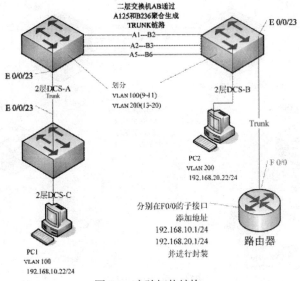

图 3-2　实验拓扑结构

三、实践步骤

1. 设备准备：连接单臂路由器、交换机和 PC，确保所有连接正确。
2. 网络规划：根据实验需求，规划网络拓扑，确定单臂路由器、交换机和 PC 的连接方式。

3. 设备配置：配置单臂路由器的接口和 VLAN，设置单臂路由器与交换机之间的聚合链路。
4. 测试和验证：使用 ping 命令测试网络连通性，并验证配置是否正确。
5. 故障排除：进行网络故障诊断并排除故障。
6. 文档编写：记录网络配置、变更情况及测试结果。

具体要求：如图 3-2 所示，首先在交换机 A 和交换机 B 之间启动生成树协议，经过配置，实现 PC1 与 PC2 之间的 ping 通。

实践三　动态路由 RIP-OSPF 实践项目

一、实践目的

1. 实现路由器上 RIP 和 OSPF 的配置。
2. 利用 RIP 和 OSPF 协议实现不同网络之间的互联。
3. 诊断并排除网络配置过程中出现的网络故障。

二、实践拓扑

该实验拓扑结构如图 3-3 所示。

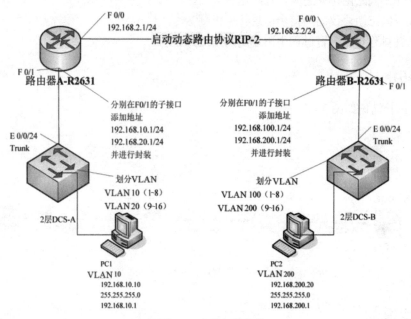

图 3-3　实验拓扑结构

三、实践步骤

1. 设备准备：连接路由器、交换机和计算机，确保所有设备连接正确。
2. 网络规划：根据实验需求规划网络拓扑，确定路由器、交换机和计算机的连接方式。
3. 设备配置：配置路由器的接口和 VLAN，设置 RIP 和 OSPF，包括宣告网络、路由器 ID 配置和区域设置。
4. 测试和验证：使用 ping 命令测试网络连通性，验证配置的正确性。
5. 故障排除：进行网络故障诊断并排除故障。
6. 文档编写：记录网络配置、变更情况及测试结果。

具体要求：参照图 3-3 所示，在交换机 A 和交换机 B 划分对应 VLAN，并在路由器 A 和路由器 B 上进行相应配置。配置动态路由 RIP-2 后，使 PC1 与 PC2 之间能够 ping 通。

实践四　路由与多层交换机间的 RIP-OSPF-静态路由 重点分布实践项目

一、实践目的

1. 理解并掌握路由协议 RIP、OSPF 及静态路由的配置方法。
2. 学会在不同路由协议之间进行路由重分布。
3. 掌握不同路由协议间的路由信息传递机制。
4. 诊断并排除网络配置过程中出现的网络故障。

二、实践拓扑

该实验拓扑结构如图 3-4 所示。

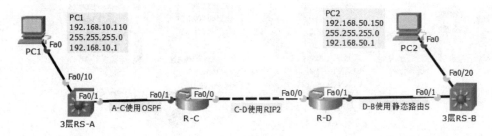

图 3-4　实验拓扑结构

三、实践步骤

1. 设备准备：根据实验要求，对路由器、多层交换机以及计算机按照实验要求进行物理连接，并确保所有设备之间的连接无误。

2. 网络规划：根据实验目的设计网络拓扑结构，明确每台路由器、交换机和计算机的连接关系。并确定 VLAN 划分、接口 IP 地址分配以及路由协议的配置细节。

3. 设备配置：

(1) 路由器接口配置。为路由器的各个接口配置 IP 地址，并确保接口状态正常。

(2) VLAN 配置。在交换机上创建 VLAN，并将相应的接口加入到对应的 VLAN 中。

(3) 路由器与交换机聚合链路配置。在路由器和交换机之间配置聚合链路(如 EtherChannel)，以提高链路带宽和冗余性。

(4) 路由协议配置。按照实验要求配置 RIP、OSPF 和静态路由，并设置路由重分布。

4. 测试和验证：使用 ping 命令测试网络连通性，以验证配置的正确性。

5. 故障排除：进行网络故障诊断并排除故障。

6. 文档编写：记录实验中的网络配置命令、接口 IP 分配、VLAN 设置、聚合链路配置等详细信息；记录测试过程和结果，包括 ping 测试的连通性报告。编写完整的实验报告，包括实验目的、步骤、结果及分析。

具体要求：参照图 3-4 所示进行设备连线，并对各个设备进行以下配置。

(1) 三层交换机 A：
    划分   VLAN 100(9~16)，   VLAN 200 (1)；
            Intvl100   192.168.10.1/24；
            Intvl200   192.168.20.1/30；

(2) 设置路由器 C：
    设置   F 0/1   192.168.20.2/30；
            F 0/0   192.168.30.2/30

(3) 设置路由器 D：
    设置   F 0/0   192.168.30.1/30
            F 0/1   192.168.40.2/30

(4) 三层交换机 B：
    划分   VLAN 400(17~24)，   VLAN 200(1)；
            Intvl200   192.168.40.1/30；
            Intvl400   192.168.50.1/24；

(5) PC1 设置：   PC2 设置：
    192.168.10.110   192.168.50.150
    255.255.255.0   255.255.255.0
    192.168.10.1   192.168.50.1

(6) 根据连线位置正确配置 PC1 和 PC2 的地址(包括网关地址以及与 VLAN 的对应关系)。

实验结果：PC1---ping---PC2   通。

查看交换机状态：sh vlan   sh run   sh ip route。

主要配置命令参考

启动 SA 到 RC 的动态路由使用 OSPF 协议。根据不同版本的三层交换机，命令可能有所不同(如果以下方法 1 的命令不能使用时，可以尝试使用方法 2)。

方法 1：

```
SA(config)#int vl 100
SA(config)# ip ospf enable area 0
SA(config)#int vl 200
SA(config)# ip ospf enable area 0
SA(config)#router ospf
SA(config-Router-ospf)#redistribute ospfase connected
SA(config-Router-ospf)#redistribute ospfase static
SA(config-Router-ospf)#redistribute ospfase rip
```

方法 2：

```
SA(config)#router ospf
SA(config-Router)#network 192.168.10.0 255.255.255.0 area 0
SA(config-Router)#network 192.168.20.0 255.255.255.252 area 0
SA(config-Router)#redistribute   connected
SA(config-Router)#redistribute   static
SA(config-Router)#redistribute   rip

RC_config#router ospf   1
RC_config_Router#network 192.168.20.0 255.255.255.252 area 0
RC_config_Router# redistribute connect
RC_config_Router# redistribute static
RC_config_Router# redistribute rip
```

启动 RC---RD 的动态路由 RIP 协议：

```
RC_config#router rip
RC_config_rip#version 2
RC_config_rip#network 192.168.30.0
RC_config_rip#auto-summary                    ----自动汇总
RC_config_rip#redistribute ospf   1           ----转发 OSPF 路由
RC_config_rip#redistribute connect            ----转发直连路由
RC_config_rip#redistribute static             ----转发静态路由
RD_config#router rip
RD_config_rip#version 2
RD_config_rip#network 192.168.30.0
RD_config_rip#auto-summary
RD_config_rip#redistribute connect
RD_config_rip#redistribute static
```

启动 RD---SB 的静态路由协议：

```
RD_config#ip route    192.168.50.0 255.255.255.0 192.168.40.1
SB(config)#ip route   192.168.30.0 255.255.255.252 192.168.40.2
SB(config)#ip route   192.168.20.0 255.255.255.252 192.168.40.2
SB(config)#ip route   192.168.10.0 255.255.255.0 192.168.40.2
```

验证：sh run   sh ip route。

验证命令：ping。

PC1 运行 cmd。

1. ping 192.168.10.1   -t   通
2. ping 192.168.20.1   -t   通
3. ping 192.168.30.1   -t   通
4. ping 192.168.40.1   -t   通
5. ping 192.168.50.1   -t   通
6. ping 192.168.50.150 -t   通

PC2 运行 cmd。

1. ping 192.168.10.1   -t   通
2. ping 192.168.20.1   -t   通
3. ping 192.168.30.1   -t   通
4. ping 192.168.40.1   -t   通
5. ping 192.168.50.1   -t   通
6. ping 192.168.10.110 -t   通。

实践五　路由与多层交换机间的静态路由-OSPF-RIP 实践项目

一、实践目的

1. 理解并掌握路由协议 RIP、OSPF 及静态路由的配置方法。
2. 学会在不同的路由协议之间进行路由重分布。
3. 掌握不同网络路由协议之间的路由信息传递机制。
4. 诊断并排除网络配置过程中出现的故障。

二、实践拓扑

该实验拓扑结构如图 3-5 所示。

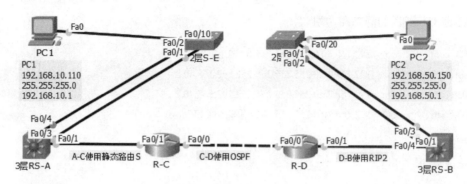

图 3-5 实验拓扑结构

三、实践步骤

1. 设备准备：根据实验要求，将路由器、多层交换机和计算机进行物理连接，确保所有设备之间的连接正确无误。

2. 网络规划：根据实验目的设计网络拓扑结构，明确每台路由器、交换机和计算机的连接关系。确定 VLAN 划分、接口 IP 地址分配以及路由协议的配置细节。

3. 设备配置：参照以下步骤配置。

(1) 路由器接口配置。为路由器的各个接口配置 IP 地址，并确保接口状态正常。

(2) VLAN 配置。在交换机上创建 VLAN，并将相应的接口分配到各个 VLAN 中。

(3) 路由器与交换机聚合链路配置。在路由器和交换机之间配置聚合链路(如 EtherChannel)，以提高链路带宽和冗余性。

(4) 路由协议配置。根据实验要求配置 RIP、OSPF 和静态路由，并设置路由重分布。

4. 测试和验证：使用 ping 命令测试网络连通性，以验证配置是否正确。

5. 故障排除：进行网络故障诊断和排除。

6. 文档编写：记录实验中的网络配置命令、接口 IP 分配、VLAN 设置以及聚合链路配置等详细信息。记录测试过程和结果，包括 ping 测试的连通性报告。编写完整的实验报告，包括实验目的、步骤、结果和分析。

具体要求：参照图 3-5 所示进行连线，各设备的配置如下。

(1) 三层交换机 A：
　　划分　　VLAN 100(9~16)，　　VLAN 200 (1)，trunk(3~4)；
　　　　　　Intvl100　　　　　　192.168.10.1/24；
　　　　　　Intvl200　　　　　　192.168.20.1/30；

(2) 设置路由器 C：
　　设置　　F 0/1　　　　　　　192.168.20.2/30；
　　　　　　F 0/0　　　　　　　192.168.30.2/30

(3) 设置路由器 D：
　　设置　　F 0/0　　　　　　　192.168.30.1/30
　　　　　　F 0/1　　　　　　　192.168.40.2/30；

(4) 三层交换机 B：
　　划分　　VLAN 400(17~24),　　VLAN 200(1)，Trunk(3-4);
　　　　Intvl200　　　　　　192.168.40.1/30;
　　　　Intvl400　　　　　　192.168.50.1/24;
(5) 二层交换机 E：
　　划分 VLAN 100 (9~16)，VLAN 200(17~24)，Trunk(1~2)。
　　E12 与 A34 生成聚合的 Trunk 链路。
(6) 二层交换机 F：
　　划分 VLAN 400 (9~16)，VLAN 200(17~24)，Trunk(1~2)。
　　F12 与 B34 生成聚合的 Trunk 链路。
(7) 根据连线位置正确配置 PC1 和 PC2 的地址(包括网关地址以及与 VLAN 的对应关系)。

实验结果：PC1---ping---PC2　　　通。
查看交换机状态：sh vlan　　　sh run　　　sh ip route。

实践六　特定网络综合设计与实现实践项目

一、实践目的

1. 理解和掌握路由协议 RIP、OSPF 以及静态路由的配置方法。
2. 学会在不同的路由协议之间进行路由重分布。
3. 掌握不同路由协议之间的路由信息传递机制。
4. 诊断并排除网络配置过程中出现的故障。

二、实践拓扑

该实验拓扑结构如图 3-6 所示。

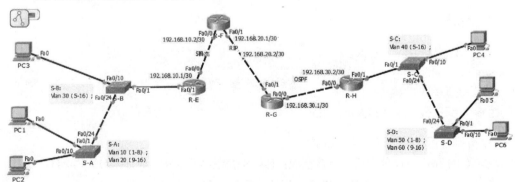

图 3-6　实验拓扑结构

三、实践步骤

1. 设备准备：根据实验要求，将路由器、多层交换机以及 PC 进行物理连接，确保所有设备之间的连接正确无误。

2. 网络规划：根据实验目的，设计网络拓扑结构，明确每台路由器、交换机和 PC 的连接关系。确定 VLAN 划分、接口 IP 地址分配以及路由协议的配置细节。

3. 设备配置：参照以下步骤配置。

(1) 路由器接口配置。为路由器的各个接口配置 IP 地址，并确保接口状态正常。

(2) VLAN 配置。在交换机上创建 VLAN，并将相应的接口加入到 VLAN 中。

(3) 路由器与交换机聚合链路配置。在路由器和交换机之间配置聚合链路(如 EtherChannel)，以提高链路带宽和冗余性。

(4) 路由协议配置。按照实验要求配置 RIP、OSPF 和静态路由，并设置路由重分布。

4. 测试和验证：使用 ping 命令测试网络连通性，以验证配置是否正确。

5. 故障排除：进行网络故障诊断和排除。

6. 文档编写：记录实验中的网络配置命令、接口 IP 分配、VLAN 设置、聚合链路配置等详细信息。记录测试过程和结果，包括 ping 测试的连通性报告。编写完整的实验报告，包括实验目的、步骤、结果和分析。

四、实验要求

参照图 3-6 所示进行连线，并配置各个设备。

1. 设备 IP 地址、gateway、Mask(连线-直通线)：

```
PC1 192.168.1.101 192.168.1.1 255.255.255.0 S-A:  0/1
PC2 192.168.2.102 192.168.2.1 255.255.255.0 S-A:  0/10
PC3 192.168.3.103 192.168.3.1 255.255.255.0 S-B:  0/10
PC4 192.168.4.104 192.168.4.1 255.255.255.0 S-C:  0/10
PC5 192.168.5.105 192.168.5.1 255.255.255.0 S-D:  0/1
PC6 192.168.6.106 192.168.6.1 255.255.255.0 S-D:  0/10
```

2. 交换机设置：

```
S-A:  VLAN 10(1~8);  VLAN 20(9~16)
S-B:  VLAN 30(5~16);
S-C:  VLAN 40(5~16);
S-D:  VLAN 50(1~8);  VLAN 60(9~16)
```

3. 分别使用 3 种协议设置路由器。

E-F-G-H 路由协议分别适用链接情况：E----S 静态----F----RIP----G----OSPF----H。

使得：PC1----ping-----PC2; -PC3; -PC4; -PC5; -PC6 全通。

4. 查看交换机状态：sh vlan   sh run   sh ip route。

附 录

一、交换机相关配置命令详解

show running-config

命令：show running-config
功能：显示当前运行状态下的交换机配置，包括所有生效的参数设置。
命令模式：特权用户配置模式。
默认配置：对于当前生效的配置参数，如果与默认参数相同，则不显示。

show version

命令：show version
功能：显示交换机版本信息。
命令模式：特权用户配置模式。

config

命令：config [terminal]
功能：从特权用户配置模式进入到全局配置模式。
参数：[terminal]表示进行终端配置。
命令模式：特权用户配置模式。

enable

命令：enable
功能：使用 enable 命令，将用户从一般用户配置模式切换到特权用户配置模式。
命令模式：一般用户配置模式。

enable password

命令：enable password
功能：修改从一般用户配置模式进入特权用户配置模式的口令，输入本命令后，系统将提示输入<Current password>和<New password>参数，用户需要进行配置。
参数：<Current password>为原有密码，最长不超过 16 个字符；<New password>为新密码，最长不超过 16 个字符；<Confirm new password>为确认新密码，值应与新密码相同，否则需要重新设置密码。

命令模式：全局配置模式。

默认配置：系统默认的特权用户密码为空。在首次配置时，当系统提示输入原密码时，直接按回车键即可。

exit

命令：exit

功能：从当前模式退出，返回到上一个模式。例如在，全局配置模式使用此命令返回到特权用户配置模式，在特权用户配置模式下使用此命令退回到一般用户配置模式等。

命令模式：各种配置模式。

help

命令：help

功能：输出关于命令解释器帮助系统的简要说明。

命令模式：各种配置模式。

clock set

命令：clock set <HH:MM:SS> <YYYY/MM/DD>

功能：设置系统日期和时钟。

参数：<HH:MM:SS >为当前时钟，HH 的取值范围为 0~23，MM 和 SS 取值范围为 0~59；YYYY/MM/DD >为当前年、月和日，YYYY 的取值范围为 2000~2100，MM 的取值范围为 1~12，DD 的取值范围为 1~31。

命令模式：特权用户配置模式。

默认配置：系统启动时默认时间为 2001 年 1 月 1 日 00：00：00。

hostname

命令：hostname <hostname>

功能：设置交换机命令行界面的提示符。

参数：<hostname>为提示符的字符串。

命令模式：全局配置模式。

默认配置：系统默认提示符为"DCS-3926S"。

language

命令：language {chinese|english}

功能：设置显示的帮助信息语言类型。

参数：chinese 为中文显示；english 为英文显示。

命令模式：特权用户配置模式。

reload

命令：reload
功能：热启动交换机。
命令模式：特权用户配置模式。
使用指南：用户可以通过此命令在不关闭电源的情况下重启交换机。

set default

命令：set default
功能：恢复交换机的出厂设置。
命令模式：特权用户配置模式。
使用指南：恢复交换机的出厂设置，即用户对交换机做的所有配置将被清除。重启交换机后，显示的提示信息将与交换机首次上电时相同。
注意：执行此命令后，必须执行 write 命令进行配置保存，然后重启交换机，以使交换机恢复到出厂设置。

show flash

命令：show flash
功能：显示保存在 Flash 中的文件及其大小。
命令模式：特权用户配置模式。

show startup-config

命令：show startup-config
功能：显示当前运行状态下存储在 Flash Memory 中的交换机参数配置。该配置通常是交换机下次上电启动时所使用的配置文件。
命令模式：特权用户配置模式。
默认配置：如果从 Flash 中读取的配置参数与默认工作参数相同，则不显示。

ip address

命令：ip address <ip-address> <mask> [secondary]
　　　no ip address [<ip-address> <mask>] [secondary]
功能：设置交换机的 IP 地址及子网掩码；使用 no 命令可以删除该 IP 地址配置。
参数：<ip-address>以点分十进制格式表示的 IP 地址；<mask>以点分十进制格式表示的子网掩码，[secondary]标识配置的 IP 地址为辅助 IP 地址。
命令模式：VLAN 接口配置模式。
默认配置：出厂时交换机未配置 IP 地址。
相关命令：ip bootp-client enable、ip dhcp-client enable

interface vlan

命令：interface vlan <vlan-id>

　　　no interface vlan <vlan-id>

功能：创建一个 VLAN 接口，即创建一个交换机的三层接口；使用 no 命令可删除交换机的三层接口。

参数：<vlan-id>是已建立的 VLAN 的 VLAN ID。

命令模式：全局配置模式。

默认配置：出厂时没有配置三层接口。

ping

命令：ping [<ip-addr>]

功能：交换机向远端设备发送 ICMP 请求包，以检测交换机与远端设备之间的可达性。

参数：<ip-addr>为要 ping 的目标主机的 IP 地址，采用点分十进制格式。

命令模式：特权用户配置模式。

默认配置：默认发送 5 个 ICMP 请求包；每个包的大小为 56 字节；超时时间为 2 秒。

shutdown

命令：shutdown

　　　no shutdown

功能：关闭指定的以太网端口；no 命令用于启动该端口。

命令模式：端口配置模式。

默认配置：以太网端口默认状态为开启。

telnet

命令：telnet [<ip-addr>] [<port>]

功能：以 Telnet 方式登录到 IP 地址为<ip-addr>的远程主机。

参数：<ip-addr>为远端主机的 IP 地址，采用点分十进制格式；<port>为端口号，取值范围为 0~65535。

命令模式：特权用户配置模式。

telnet-server enable

命令：telnet-server enable

　　　no telnet-server enable

功能：启用交换机的 Telnet 服务器功能；no 命令用于关闭交换机的 Telnet 服务器功能。

命令模式：全局配置模式。

默认配置：系统默认启用 Telnet 服务器功能。

telnet server securityip

命令：telnet-server securityip <ip-addr>

　　　　no telnet-server securityip <ip-addr>

功能：配置交换机作为 Telnet 服务器，允许特定安全 IP 地址的 Telnet 客户端登录；no 命令用于删除指定的 Telnet 客户端的安全 IP 地址。

参数：<ip-addr>可以访问本交换机的安全 IP 地址，采用点分十进制格式。

命令模式：全局配置模式。

默认配置：系统默认不配置任何安全 IP 地址。

telnet-user

命令：telnet-user <username> password {0|7} <password>

　　　　no telnet-user <username>

功能：设置 Telnet 客户端的用户名及密码；no 命令用于删除该 Telnet 用户。

参数：<username>为 Telnet 客户端用户名，最长不超过 16 个字符；<password>为登录密码，最长不超过 8 个字符；0 表示密码不加密显示；7 表示密码加密显示。

命令模式：全局配置模式。

默认配置：系统默认没有设置 Telnet 客户端的用户名及密码。

ip http server

命令：ip http server

　　　　no ip http server

功能：启用 Web 配置功能；no 操作用于关闭 Web 配置。

命令模式：全局配置模式。

web-user

命令：web-user <username> password {0|7} <password>

　　　　no web-user <username>

功能：设置 Web 客户端的用户名及密码；no 命令用于删除该 Web 用户。

参数：<username>为 Web 访问的授权用户名，最长不超过 16 个字符；<password>为登录密码，最长不超过 8 个字符；0 表示密码不加密显示；7 表示密码加密显示。

命令模式：全局配置模式。

no name

命令：name <vlan-name>

功能：为 VLAN 指定名称，VLAN 的名称是对该 VLAN 的描述性字符串；no 命令用于删除 VLAN 的名称。

参数：<vlan-name>为指定的 VLAN 名称字符串。

命令模式：VLAN 配置模式。
默认配置：VLAN 默认名称为 VLAN XXX，其中 XXX 为 VLAN ID(VID)。

vlan

命令：vlan <vlan-id>
　　　　no vlan <vlan-id>

功能：创建 VLAN 并进入 VLAN 配置模式；在 VLAN 模式中，用户可以配置 VLAN 名称并为该 VLAN 分配交换机端口；no 命令用于删除指定的 VLAN。

参数：<vlan-id>为要创建/删除的 VLAN 的 VID，取值范围为 1~4094。

命令模式：全局配置模式。

默认配置：交换机默认仅包含 VLAN 1。

switchport interface

命令：switchport interface <interface-list>
　　　　no switchport interface <interface-list>

功能：为 VLAN 分配以太网端口；no 命令用于删除指定 VLAN 内的一个或一组端口。

参数：<interface-list>为要添加或者删除的端口列表，支持使用分号(;)和连字符(-)，例如：ethernet 0/0/1;2;5 或 ethernet 0/0/1-6;8。

命令模式：VLAN 配置模式。

默认配置：新建立的 VLAN 默认不包含任何端口。

switchport mode

命令：switchport mode {trunk|access}

功能：设置交换机的端口为 Access 模式或者 Trunk 模式。

参数：trunk 表示端口允许通过多个 VLAN 的流量；Access 表示端口只能属于一个 VLAN。

命令模式：端口配置模式。

默认配置：端口默认为 Access 模式。

switchport access vlan

命令：switchport access vlan <vlan-id>
　　　　no switchport access vlan

功能：将当前 Access 端口加入指定 VLAN；no 命令用于将当前端口从 VLAN 中删除。

参数：<vlan-id>为当前端口要加入的 VLANVID，取值范围为 1~4094。

命令模式：端口配置模式。

默认配置：所有端口默认属于 VLAN 1。

switchport mode

命令：switchport mode {trunk|access}
功能：设置交换机的端口为 Access 模式或者 Trunk 模式。
参数：trunk 表示端口允许通过多个 VLAN 的流量；access 表示端口只能属于一个 VLAN。
命令模式：端口配置模式。
默认配置：端口默认为 Access 模式。

switchport trunk allowed vlan

命令：switchport trunk allowed vlan {<vlan-list>|all}
　　　no switchport trunk allowed vlan
功能：设置 Trunk 端口允许通过 VLAN；no 命令用于为恢复默认情况。
参数：<vlan-list>为允许在该 Trunk 端口上通过的 VLAN 列表；all 表示允许该 Trunk 端口通过所有 VLAN 的流量。
命令模式：端口配置模式。
默认配置：Trunk 端口默认允许通过所有 VLAN。

switchport trunk native vlan

命令：switchport trunk native vlan <vlan-id>
　　　no switchport trunk native vlan
功能：设置 Trunk 端口的 PVID；no 命令用于恢复默认值。
参数：<vlan-id>为 Trunk 端口的 PVID。
命令模式：端口配置模式。
默认配置：Trunk 端口默认的 PVID 为 1。

am enable

命令：am enable
　　　no am enable
功能：启用访问控制功能。在启用时，AM 模块将拒绝所有 IP 报文通过；no 命令则禁止访问管理功能，并清空 IP 地址池和 MAC-IP 地址池。
命令模式：全局配置模式。
默认配置：AM 访问控制默认未启用。

am ip_pool

命令：am ip-pool <start_ip_address> [<num>]
　　　no am ip-pool<start_ip_address> [<num>]
功能：创建一个 IP 地址段并添加到地址池中；no 命令用于删除已配置的 IP 地址段。
参数：start_ip_address 是 IP 地址范围的起始地址；num 是从 start_ip_address 开始的连

续的地址数量，默认值为1。

命令模式：物理接口配置模式。

默认配置：IP池默认为空。

am mac_ip_pool

命令：am mac-ip-pool <mac_address> <ip_address>
　　　no am mac-ip-pool <mac_address>< ip_address>

功能：创建一个MAC与IP地址的绑定并添加到地址池中，或删除配置的MAC与IP地址的绑定，确保MAC和IP地址一一对应。

参数：

(1) mac_address为源MAC地址，格式为HH-HH-HH-HH-HH-HH。

(2) ip_address为源IP地址，32位二进制数，表示为四个用十进制数隔开的部分。

命令模式：物理接口配置模式。

默认配置：MAC与IP池默认为空。

am port

命令：am port
　　　no am port

功能：打开或关闭物理接口上的AM功能。

命令模式：物理接口配置模式。

默认配置：AM功能默认处于开启状态。

no am all

命令：no am all {ip-pool|mac-ip-pool}

功能：删除所有用户配置的MAC-IP地址池或IP地址池。

ip-pool：表示IP地址池。

mac-ip-pool：MAC-IP地址池。

all：表示所有IP地址池或MAC地址池。

命令模式：全局配置模式。

默认配置：无。

show am

命令：show am [interface <interfaceName>]

功能：显示当前交换机配置的地址项。

Interface Name：指定物理接口名称。

命令模式：全局配置模式。

默认配置：无。

spanning-tree

命令：spanning-tree

no spanning-tree

功能：在交换机的全局配置模式和端口配置模式下启用或禁用 MSTP 协议。no 操作用于关闭 MSTP 协议。

参数：无。

命令模式：全局配置模式和端口配置模式。

默认配置：系统默认不运行 MSTP 协议。但是，在全局配置模式下启用 MSTP 协议后，所有的端口默认会开启 MSTP 协议。

port-group

命令：port-group <port-group-number> [load-balance { src-mac|dst-mac | dst-src-mac | src-ip|dst-ip|dst-src-ip}]

no port-group <port-group-number> [load-balance]

功能：新建一个端口组，并且设置该组的流量分配方式。如果未指定流量分担方式，则使用默认的流量分担方式。no 命令可以删除该端口组或恢复该组流量分担的默认值。指定 load-balance 表示恢复默认流量分担方式，否则为删除该组。

参数：<port-group-number>为 Port Channel 的组号，范围为 1~16。如果已经存在该组号则会报错。dst-mac 根据目的 MAC 进行流量分担。src-mac 根据源 MAC 地址进行流量分担；dst-src-mac 根据目的 MAC 和源 MAC 进行流量分担；dst-ip 根据目的 IP 地址进行流量分担。src-ip 根据源 IP 地址进行流量分担。dst-src-ip 根据目的 IP 和源 IP 地址进行流量分担。如果修改流量分担方式，并且该端口组已经形成一个端口通道，则这次修改的流量分担方式将在下次汇聚时生效。

命令模式：交换机全局配置模式。

默认配置：默认交换机端口不属于 Port Channel，且不启动 LACP 协议。

port-group mode

命令：port-group <port-group-number> mode {active|passive|on}

no port-group <port-group-number>

功能：将物理端口加入 Port Channel。使用 no 命令可以将端口从 Port Channel 中移除。

参数：<port-group-number>为 Port Channel 的组号，范围为 1~16；active(0)启动端口的 LACP 协议，并设置为 Active 模式；passive(1)启动端口的 LACP 协议，并且设置为 Passive 模式；on 强制端口加入 Port Channel，且不启动 LACP 协议。

命令模式：交换机端口配置模式。

默认配置：默认交换机端口不属于 Port Channel，且不启动 LACP 协议。

interface port-channel

命令：interface port-channel <port-channel-number>
功能：进入汇聚端口配置模式。
命令模式：全局配置模式。

show port-group

命令：show port-group [<port-group-number>] {brief | detail | load-balance | port | port-channel}
参数：<port-group-number>为要显示的 Port Channel 的组号，范围为 1~16；brief 显示摘要信息；detail 显示详细信息；load-balance 显示流量分担信息；port 显示成员端口信息；port-channel 显示汇聚端口信息。
命令模式：特权配置模式。

exec timeout

命令：exec timeout <minutes>
功能：设置退出特权用户配置模式的超时时间。
参数：<minute>为时间值，单位为分钟，取值范围为 0~300。
命令模式：全局配置模式。
默认配置：系统默认的超时时间为 5 分钟。
使用指南：为确保交换机使用的安全性，防止非法用户的恶意操作，当特权用户在完成最后一项配置后，开始计时。到达设置时间值时，系统将自动退出特权用户配置模式。如果用户想再次进入特权用户配置模式，需要再次输入特权用户密码。如果将 exec timeout 的值设置为 0，则系统不会退出特权用户配置模式。
举例：设置交换机退出特权用户配置模式的超时时间为 6 分钟。

switch(config)#exec timeout 6

命令：ip host <hostname> <ip_addr>
　　　no ip host <hostname>
功能：设置主机名称与 IP 地址的映射关系。no 命令用于删除该项映射关系。
参数：<hostname>为主机名称，最长不超过 15 个字符；<ip_addr>为主机名对应 IP 地址，采用点分十进制格式。
命令模式：全局配置模式。

auto-summary

命令：auto-summary
　　　no auto-summary
功能：配置路由聚合功能。使用 no 命令可以取消路由聚合功能。

参数：无。
命令模式：RIP 协议配置模式。
默认配置：默认不使用自动路由聚合功能。

default-metric

命令：default-metric <value>
　　　no default-metric
功能：设定引入路由的默认路由权值。使用 no 命令可以恢复默认设置。
参数：<value>为所要设定的路由权值，取值范围为 1~16。
命令模式：RIP 协议配置模式。
默认配置：默认的路由权值为 1。

ip rip authentication key-chain

命令：ip rip authentication key-chain <name-of-chain>
　　　no ip rip authentication key-chain
功能：设置 RIP 验证使用的密钥。使用 no 命令可以取消对 RIP 验证的设置。
参数：<name-of-chain>字符串，最长不超过 16 个字符。
命令模式：接口配置模式。
默认配置：系统默认不进行 RIP 验证。

ip rip authentication mode

命令：ip rip authentiaction mode {text|md5 type {cisco|usual}}
　　　no ip rip authentication mode
功能：设置验证所使用的类型。使用 no 命令可以恢复默认验证类型，即文本验证。
参数：text 表示使用文本验证；md5 表示 MD5 验证，并且 MD5 验证又分为 Cisco MD5 和常规 MD5 两种验证方法。
命令模式：接口配置模式。
默认配置：默认使用文本验证。

ip rip metricin

命令：ip rip metricin <value>
　　　no ip rip metricin
功能：设置在接口接收 RIP 报文增加的附加路由权值。使用 no 命令可以恢复默认设置。
参数：<value>为附加的路由权值，取值范围为 1~15。
命令模式：接口配置模式。
默认配置：RIP 在接收报文时的默认附加路由权值为 1。
相关命令：ip rip metricout

ip rip metricout

命令：ip rip metricout <value>
　　　　no ip rip metricout
功能：设置从接口发送 RIP 报文时增加的附加路由权值。使用 no 命令可以为恢复默认设置。
参数：<value>为附加的路由权值，取值范围为 0~15。
命令模式：接口配置模式。
默认配置：RIP 在发送报文时的默认附加路由权值为 0。

ip rip input

命令：ip rip input
　　　　no ip rip input
功能：设置接口是否能够接收 RIP 报文。使用 no 命令可以禁用接口接收 RIP 报文。
命令模式：接口配置模式。
默认配置：接口默认能够接收 RIP 报文。

ip rip output

命令：ip rip output
　　　　no ip rip output
功能：设置接口是否能够向外发送 RIP 报文。使用 no 命令可以禁用接口向外发送 RIP 报文。
命令模式：接口配置模式。
默认配置：接口默认能够向外发送 RIP 报文。

ip rip receive version

命令：ip rip receive version {v1 | v2 | v12}
　　　　no ip rip receive version
功能：设置接口接收 RIP 报文的版本信息(默认接收 RIP 版本 1 和 2)。使用 no 命令可恢复到默认设置。
参数：v1 和 v2 表示接收 RIP 版本 1 和 RIP 版本 2 报文；v12 表示接收 RIP 版本 1 和版本 2 报文。
命令模式：接口配置模式。
默认配置：默认接收 RIP 版本 1 和版本 2 报文。

ip rip send version

命令：ip rip send version { v1 | v2 [bc|mc] }
　　　　no ip rip send version
功能：设置接口发送 RIP 报文的版本。使用 no 命令可以恢复默认设置。
参数：v1 和 v2 为发送 RIP 版本 1 和版本 2 报文。[bc|mc]只在发送 RIP 版本 2 的情况

下配置，用于指定发送方式，BC 为广播方式，MC 为组播方式。当配置发送 RIP 版本 2 报文时，接口默认以组播 MC 方式发送 RIP 版本 2 报文，只有在设置 BC 后才能在此接口发送广播报文。

命令模式：接口配置模式。

默认配置：接口默认发送 RIP 版本 2 报文。

ip rip work

命令：ip rip work

　　　no ip rip work

功能：设置接口上是否运行 RIP 协议。使用 no 命令可以使接口不再收发 RIP 报文。

命令模式：接口配置模式。

默认配置：在启用 RIP 路由功能后，接口默认会运行 RIP 协议。

ip split-horizon

命令：ip split-horizon

　　　no ip split-horizon

功能：设置是否允许水平分割。使用 no 命令可以禁止水平分割。

命令模式：接口配置模式。

默认配置：默认情况下，允许水平分割。

redistribute

命令：redistribute { static | ospf | bgp } [metric <value>]

　　　no redistribute { static | ospf | bgp }

功能：在 RIP 路由中引入其他路由协议的路由。使用 no 命令可以取消引入。

参数：static 指定引入静态路由；ospf 指定引入 OSPF 路由；bgp 指定引入 BGP 路由；<value>指定以多大的路由权值引入路由，取值范围 1~16。

命令模式：RIP 配置模式。

默认配置：RIP 默认不引入其他路由。如果引入其他的路由协议而未指定其权值，则按默认路由权值 default-metric 引入。

rip broadcast

命令：rip broadcast

　　　no rip broadcast

功能：配置三层交换机的所有接口是否发送 RIP 广播包或组播包。使用 no 命令可以禁止各端口发送广播包或组播包，仅允许在配置了 neighbor 的三层交换机之间发送 RIP 数据包。

命令模式：RIP 配置模式。

默认配置：系统默认发送 RIP 广播包。

rip checkzero

命令：rip checkzero
　　　no rip checkzero
功能：使用本命令对 RIPI 协议报文中的零域进行检查。使用 no 命令可以取消对零域的查零操作。由于 RIPv2 的报文中没有零域，所以本命令对 RIPv2 没有作用。
命令模式：RIP 协议配置模式。
默认配置：系统默认对 RIPv1 报文进行零域检查。

rip preference

命令：rip preference <value>
　　　no rip preference
功能：指定 RIP 协议的路由优先级；使用 no 命令可以恢复默认值。
参数：<value>指定优先级的值，取值范围为 0~255。
命令模式：RIP 协议配置模式。
默认配置：系统默认将 RIP 的优先级设置为 120。

router rip

命令：router rip
　　　no router rip
功能：开启 RIP 路由进程并进入 RIP 配置模式。使用 no 命令可以关闭 RIP 路由协议。
命令模式：全局配置模式。
默认配置：不运行 RIP 路由协议。

timer basic

命令：timer basic <update> <invalid> <holddown>
　　　no timer basic
功能：调整 RIP 计时器更新间隔、无效时间和抑制时间。使用 no 命令可以恢复各项参数的默认值。
参数：<update>发送更新报文的时间间隔，单位秒，取值范围 1~2 147 483 647；<invalid>宣布 RIP 路由无效的时间段，单位秒，取值范围 1~2 147 483 647；<holddown>为宣布某路由无效后仍可在路由表中存在的时间段，单位秒，取值范围 1~2 147 483 647。
命令模式：RIP 协议配置模式。
默认配置：<update>默认值为 30；<invalid>默认值为 180；<holddown>默认值为 120。

version

命令：version {1|2}
　　　no version

功能：设置所有路由器接口发送和接收 RIP 数据包的版本。使用 no 命令可以恢复到默认设置(1 为 RIP 版本 1，2 为 RIP 版本 2)。

命令模式：RIP 协议配置模式。

默认配置：发送 RIP 版本 1 的数据包，并接收版本 1 和 2 的数据包。

show ip protocols

命令：show ip protocols
功能：显示三层交换机当前运行的路由协议的信息。
命令模式：特权用户配置模式。

show ip route

命令：show ip route
功能：显示路由表中与 RIP 路由相关的目的 IP 地址、网络掩码以及下一跳 IP 地址或转发接口等信息。

show ip protocols

命令：show ip protocols
功能：显示当前三层交换机运行路由协议的信息。

default redistribute cost

命令：default redistribute cost <cost>
　　　　no default redistribute cost
功能：配置 OSPF 在引入外部路由时所使用的默认花费值。使用 no 命令可以恢复该花费值为默认设置。

参数：<cost>代表花费值，其取值范围为 1~65 535。
命令模式：OSPF 协议配置模式。
默认配置：默认设置引入的花费值为 1。

default redistribute interval

命令：default redistribute interval <time>
　　　　no default redistribute interval
功能：配置 OSPF 引入外部路由的时间间隔。使用 no 命令可以将该时间间隔恢复为默认值。
参数：<time>表示引入外部路由的时间间隔，单位为秒，取值范围为 1~65 535。
命令模式：OSPF 协议配置模式。
默认配置：OSPF 引入外部路由的时间间隔默认为 1 秒。

default redistribute limit

命令：default redistribute limit <routes>

no default redistribute limit

功能：配置 OSPF 一次可引入外部路由的最大值。使用 no 命令可以将该最大值恢复为默认设置。

参数：<routes>表示引入路由数量的最大值，取值范围为 1~65 535。

命令模式：OSPF 协议配置模式。

默认配置：OSPF 引入外部路由数量的最大值默认设置为 100。

default redistribute tag

命令：default redistribute tag <tag>

　　　no default redistribute tag

功能：配置引入外部路由时的默认标记值。使用 no 命令可以将标记值恢复为默认设置。

参数：<tag>为标记值，取值范围为 0~4 294 967 295。

命令模式：OSPF 协议配置模式。

默认配置：默认标记值为 0。

default redistribute type

命令：default redistribute type { 1 | 2 }

　　　no default redistribute type

功能：配置引入外部路由时的默认路由类型。使用 no 命令可以将路由类型恢复为默认设置。

参数：1 表示第一类外部路由，2 表示第二类外部路由。

命令模式：OSPF 协议配置模式。

默认配置：系统默认引入的外部路由类型为第二类外部路由。

ip ospf authentication

命令：ip ospf authentication { simple <auth_key>| md5 <auth_key> <key_id>}

　　　no ip ospf authentication

功能：指定接口上接收 OSPF 报文所需要的验证方式。使用 no 命令可以取消验证设置。

参数：simple 为简单验证方式；md5 为 MD5 加密验证方式；<auth_key>为验证密钥，MD5 验证方式下最大长度为 16 字节；

<key_id>为 MD5 验证方式时的验证字，取值范围为 1~255。

命令模式：接口配置模式。

默认配置：接口上接收 OSPF 报文默认不需要验证。

ip ospf cost

命令：ip ospf cost <cost>

　　　no ip ospf cost

功能：指定接口运行 OSPF 协议所需的代价值。使用 no 命令可以恢复为默认设置。

参数：<cost >为 OSPF 协议的代价，取值范围为 1~65 535。

命令模式：接口配置模式。
默认配置：接口的 OSPF 协议代价值默认设置为 1。

ip ospf dead-interval

命令：ip ospf dead-interval <time >
   no ip ospf dead-interval
功能：指定与相邻三层交换机之间的路由失效时间长度。使用 no 命令可以恢复为默认设置。
参数：<time >为相邻三层交换机失效的时间长度，单位为秒，取值范围 1~65 535。
命令模式：接口配置模式。
默认配置：三层交换机路由失效的时间长度默认设置为 40 秒(通常是 hello-interval 的 4 倍)。
dead-interval 参数应于 hello-interval 参数一致，且至少为 hello-interval 值的 4 倍。

ospf enable area

命令：ip ospf enable area <area_id>
   no ip ospf enable area
功能：配置接口属于某个 OSPF 区域。使用 no 命令可以取消该配置。
参数：<area_id>为该接口所属区域的区域号，取值范围为 0~4 294 967 295。
命令模式：接口配置模式。
默认配置：接口默认未配置为属于任何 OSPF 区域。

ip ospf hello-interval

命令：ip ospf hello-interval <time>
   no ip ospf hello-interval
功能：指定接口上发送 HELLO 报文的时间间隔。使用 no 命令可以恢复为默认设置。
参数：<time>为发送 HELLO 报文的时间间隔，单位为秒，取值范围为 1~255。
命令模式：接口配置模式。
默认配置：接口默认发送 HELLO 报文的间隔时间为 10 秒。

ip ospf passive-interface

命令：ip ospf passive-interface
   no ip ospf passive-interface
功能：将接口设置为只接收但不发送 OSPF 报文。使用 no 命令可以取消该项配置。
命令模式：接口配置模式。
默认配置：接口默认状态下会收发 OSPF 报文。

ip ospf priority

命令：ip ospf priority <priority>
   no ip ospf priority

功能：配置接口在选举"指定三层交换机"(DR)时的优先级。使用 no 命令可以恢复为默认设置。

参数：<priority>为优先级，合法取值范围为 0~255。

命令模式：接口配置模式。

默认配置：接口在选择指定三层交换机时的默认优先级值为 1。

ip ospf retransmit-interval

命令：ip ospf retransmit-interval <time>

　　　no ip ospf retransmit-interval

功能：指定接口与邻接三层交换机之间传送链路状态宣告(LSA)时的重传间隔。使用 no 命令可以恢复为默认设置。

参数：<time>为重传间隔，单位为秒，取值范围为 1~65 535。

命令模式：接口配置模式。

默认配置：默认重传间隔为 5 秒。

ip ospf transmit-delay

命令：ip ospf tranmsit-delay <time>

　　　no ip ospf transmit-delay

功能：设置在接口上传送链路状态宣告(LSA)的时延值。使用 no 命令可以恢复为默认设置。

参数：<time>为接口上传送链路状态宣告的时延值，单位为秒，取值范围为 1~65 535。

命令模式：接口配置模式。

默认配置：接口上传送链路状态宣告的默认时延值为 1 秒。

network

命令：network <network> <mask> area <area_id> [advertise | notadvertise]

　　　no network <network> <mask> area <area_id>

功能：为三层交换机的各个网络定义所属区域。使用 no 命令可以删除该项配置。

参数：<network>和<mask>为网络 IP 地址和地址通配符位，格式为点分十进制；<area_id>为区域号，取值范围为 0~4 294 967 295；advertise | notadvertise 指定是否将该网络范围的路由摘要信息广播出去。

命令模式：OSPF 协议配置模式。

默认配置：系统默认情况下没有配置网络所属的区域。若有配置，则默认会广播摘要信息。

preference

命令：preference [ase] <preference >

　　　no preference [ase]

功能：配置 OSPF 协议在各路由协议之间的优先级，以及引入的自治系统外部路由的优先级。使用 no 命令可以恢复为默认设置。

参数：ase 表示指定引入自治系统外部路由的优先级；<preference >为优先级值，取值范围为 1~255。

命令模式：OSPF 协议配置模式。

默认配置：OSPF 协议的默认优先级为 10。引入的外部路由协议的默认优先级为 150。

redistribute ospfase

命令：redistribute ospfase { bgp |connected | static | rip } [type { 1 | 2 }] [tag <tag>] [metric<cost_value>]

　　　　no redistribute ospfase { bgp |connected | static | rip}

功能：引入 BGP 路由、直连路由、静态路由和 RIP 路由作为外部路由信息。使用 no 命令可以取消引入的外部路由信息。

参数：bgp 表示引入 BGP 路由作为外部路由信息；connected 表示引入直连路由作为外部路由信息；static 表示引入静态路由作为外部路由信息；rip 表示引入 RIP 协议发现的路由作为外部路由信息；type{1|2}指定外部路由类型，1 表示第一类外部路由，2 表示第二类外部路由；tag<tag>指定路由的标记，<tag>为路由的标记值，取值范围为 0~4 294 967 295；metric<cost_value>指定路由的权值，<cost_value>为路由的权值，取值范围为 1~16 777 215。

命令模式：OSPF 协议配置模式。

默认配置：OSPF 默认不引入外部路由。

router id

命令：router id <router_id>

　　　　no router id

功能：配置运行 OSPF 协议三层交换机的 ID 号。使用 no 命令可以取消已配置的三层交换机 ID 号。

参数：<router_id>为三层交换机的 ID 号，格式为点分十进制。

命令模式：全局配置模式。

默认配置：系统默认为不配置三层交换机的 ID 号，OSPF 协议在运行时会从各接口的 IP 地址中自动选择一个作为三层交换机的 ID 号。

router ospf

命令：router ospf

　　　　no router ospf

功能：启动 OSPF 协议，进入 OSPF 配置模式。使用 no 命令可以关闭 OSPF 协议。

命令模式：全局配置模式。

默认配置：系统默认不运行 OSPF 协议。

stub cost

命令：stub cost <cost> area <area_id >

no stub area <area_id>

功能：将指定区域定义成 STUB 区域。使用 no 命令可以取消 STUB 区域的定义。

参数：<cost>为 STUB 区域默认路由的花费值，取值范围为 1~65 535；<area_id>为 STUB 区域的区域号，取值范围为 1~4 294 967 295。

命令模式：OSPF 协议配置模式。

默认配置：系统默认未配置 STUB 区域。

二、路由器相关配置命令详解

show interface

使用 show interface 命令可以在全局配置模式下查看接口状态。

```
show interface
show interface type interface-number
show interface type slot/port (用于带有非信道化 E1 的物理接口的路由器)
show interface serial slot/port:channel-group(用于显示非信道化 E1 的物理接口)
show interface serial slot/port.subinterface-number(用于显示子接口)
```

参数：实验参数如表 A-1 所示。

表 A-1 参数说明

参数	说明
type	指定要配置的接口类型
interface-number	逻辑接口序号
slot	插槽或插卡编号
port	插槽或插卡端口编号
channel-group	范围为 0~30 的 E1 信道组号，使用 channel-group 配置命令定义
subinterface-number	范围为 1~32 767 的子接口号

默认：无。

命令模式：管理模式。

使用说明：若 show interface 命令后面不带任何参数，则会显示所有接口的详细信息。

network

使用 network 命令为 RIP 协议指定连接的网络号，使用 no network 命令则可以取消一个网络号。

```
network network-numbe <network-mask>
no network network-number <network-mask>
```

参数：实验参数如表 A-2 所示。

表 A-2 参数说明

参数	说明
Network-number	直接相连网络的网络 IP 地址
Network-mask	(可选)直接相连网络的网络 IP 地址掩码

默认：无网络被指定。

命令模式：路由配置模式。

redistribute

redistribute 命令用于将路由器从一个路由域重新分布到另一个路由域。使用 no redistribute 命令可以取消路由的重新分布。

```
redistribute protocol [process-id] [route-map map-name]
no redistribute protocol [process-id] [route-map map-name]
```

参数：实验参数如表 A-3 所示。

表 A-3 参数说明

参数	说明
protocol	路由要重新分布的源协议可以是 bgp、ospf、static[ip]、connected 或 rip 等几个关键词之一，其中： (1)关键词 static[ip]用于重新分布 IP 静态路由。当路由重新分布到 IS-IS 协议中时，可以使用这个关键词 (2)关键词 connected 是指那些在接口上 IP 激活后自动建立起来的路由。对于像 OSPF 和 IS-IS 的路由协议，这些路由会作为自治系统的外部路由被重新分布
process-id	(可选)对于 bgp 或 bigp，该参数是指 16 位数字的自助系统号。对于 OSPF，这是被重新分布路由所对应的 OSPF 进程 ID，用于标识路由进程，必须是一个非零的十进制数。对于 RIP，则不需要进程标识 process-id。
route-map	(可选)该参数告诉路由映射对那些从源协议导入到当前路由协议的路由进行过滤。如果这个参数没有给出，所有路由将重新分布。如果给出这个关键词但没有列出路由映射标记，将没有路由被导入

默认：路由重新分布处于无效状态。

- protocol：没有定义路由协议。
- process-id：没有定义进程 ID。
- route-map map-tag：如果未提供 route-map 参数，则所有路由将被重新分布。如果未指定 map-tag，则不会有路由被导入。

命令模式：路由配置模式。

router ospf

配置路由器使用 OSPF 路由协议。使用 no router ospf 命令可以禁止路由器使用 OSPF。

```
router ospf process-id
no router ospf process-id
```

参数：实验参数如表 A-4 所示。

表 A-4 参数说明

参数	说明
process-id	用于内部表示 OSPF 路由处理的参数，它是本地分配的非负整数。process-id 唯一标识一个 OSPF 路由处理过程

默认：未定义 OSPF 路由处理。
命令模式：全局配置态。

ip rip authentication

使用 ip rip authentication 接口配置命令可以为 RIP-2 数据包指定认证类型。而 no ip rip authentication 命令则用于取消对 RIP 报文的认证。

```
ip rip authentication {simple | message-digest}
no ip rip authentication
```

参数：实验参数如表 A-5 所示。

表 A-5 参数说明

参数	说明
simple	明文认证类型
message-digest	MD5 密文认证类型

默认：不认证。
命令模式：接口配置模式。

network area

将一个区域中的多个网段定义为一个网络范围，可以使用 network 命令；使用 no network area 命令可以取消网络范围。

```
network network mask area area_id [ advertise | not-advertise ]
no network network mask area area_id [ advertise | not-advertise ]
```

参数：实验参数如表 A-6 所示。

表 A-6 参数说明

参数	参数说明
network	网络 IP 地址，采用点分十进制格式
mask	掩码，采用点分十进制格式
area_id	区域号
advertise 和 notadvertise	指定是否将该网络范围的路由摘要信息广播出去

默认：系统默认没有配置网络范围。
命令模式：路由配置模式。

ip ospf password

为邻接路由配置状态密码。使用 no ip ospf password 命令取消设置。

```
ip ospf password password
no ip ospf password
```

参数：实验参数如表 A-7 所示。

表 A-7 参数说明

参数	说明
password	任何连续的 8 位字符串

默认：无密码。
命令模式：接口配置模式。

area range

在域边界进行路由汇总。使用 no area range 命令可以取消设置。

```
area area-id range address mask[ not-advertise ]
no area area-id range address mask not-advertise
no area area-id range address mask
no area area-id
```

参数：实验参数如表 A-8 所示。

表 A-8 参数说明

参数	说明
password	表示要进行路由汇总的域。可以是十进制数，也可以是一个 IP 地址
address	IP 地址
mask	IP 掩码
advertise	汇总后发布
not-advertise	汇总后不发布

默认：不起作用。
命令模式：路由配置模式。

ip nat inside source

使用 ip nat inside source 全局配置命令可以开启内部源地址的 NAT。用这个命令的 no 形式可以删除静态翻译或删除和池的动态关联，注意：动态翻译规则和静态网段翻译规则在使用时不能删除。

动态 NAT 配置命令：

ip nat inside source {list access-list-name} {interface type number | pool pool-name} [overload]
no ip nat inside source {list access-list-name} {interface type number | pool pool-name} [overload]

静态单个地址 NAT 配置命令：

ip nat inside source {static {local-ip global-ip}
no ip nat inside source {static {local-ip global-ip}

静态端口 NAT 配置命令：

ip nat inside source {static {tcp | udp local-ip local-port {global-ip | interface type number} global-port}
no ip nat inside source {static {tcp | udp local-ip local-port {global-ip | interface type number} global-port}

静态网段 NAT 配置命令：

ip nat inside source {static {network local-network global-network mask}
no ip nat inside source {static {network local-network global-network mask}

参数：实验参数如表 A-9 所示。

表 A-9　参数说明

参数	说明
List access-list-name	IP 访问列表的名字。源地址符合访问列表的报文将被用地址池中的全局地址来翻译
pool name	地址池的名字，从这个池中动态地分配全局 IP 地址
interface type number	指定网络接口
overload	(可选)使路由器对多个本地地址使用一个全局的地址。当 overload 被设置后，相同或者不同主机的多个会话将用 TCP 或 UDP 端口号来区分
static local-ip	建立一条独立的静态地址翻译；参数为分配给内部网主机的本地 IP 地址。这个地址可以自由地选择，或从 RFC 1918 中分配
local-port	设置本地 TCP/UDP 端口号，范围为 1~65535
static global-ip	建立一条独立的静态地址翻译；这个参数为内部主机建立一个外部的网络可以唯一访问的 IP 地址
global-port	设置全局 TCP/UDP 端口号，范围为 1~65 535
tcp	设置 TCP 端口翻译
udp	设置 UDP 端口翻译
network local-network	设置本地网段翻译
global-network	设置全局网段翻译
mask	设置网段翻译的网络掩码

默认：任何内部源地址的 NAT 都不存在。

命令模式：全局配置模式。